Medical Principles and Practice

International Journal of the Kuwait University Health Sciences Centre

KARGER

Printed in Switzerland
on acid-free paper by
Reinhardt Druck, Basel

Appears bimonthly:
1 volume per year
(6 issues)

Guidelines for Authors

Medical Principles and Practice

Mission Statement

MPP, the journal of the Kuwait University Health Sciences Centre aims to be a publication of international repute that serves as a medium for dissemination and exchange of scientific knowledge in the health sciences.

Objectives

1. To publish original research in the health sciences.
2. To publish peer-reviewed articles of current international interest in the form of papers, solicited reviews, case reports, short communications, conference proceedings, discussion forums and letters.

Submission

Three copies of manuscripts, written in English, with three sets of illustrations (final high-resolution pictures need only be sent after acceptance of the paper) together with a soft copy on disc, should be submitted to one of the addresses below (for details please refer to: www.karger.com/electronic_submission):

Farida Al-Awadi, PhD
Editor-in-Chief, Medical Principles and Practice
Faculty of Medicine
Kuwait University
PO Box 24923
13110 Safat (Kuwait)
Tel. +965 5340479 or 4986680
Fax +965 5333094
E-Mail mpp@hsc.edu.kw

or

Editor
Medical Principles and Practice
S. Karger AG
Allschwilerstr. 10, PO Box
CH–4009 Basel (Switzerland)

or

Editor
Medical Principles and Practice
S. Karger Publishers, Inc.
26 West Avon Road, PO Box 529
Farmington, CT 06085 (USA)

One set of the illustrations must be original. Manuscripts will not be accepted by facsimile. All manuscripts should be accompanied by a cover letter from the author responsible for correspondence for the manuscript. Manuscripts are received with the explicit understanding that they are not under simultaneous consideration by any other journal. Manuscripts must comply with the ethical standards recommended by the Helsinki Declaration. The editors realize that the use of anesthetics, analgesics and tranquilizers would defeat the purpose of some experiments. However, the use of painful or otherwise noxious stimuli must be carefully and thoroughly justified in the manuscripts. Submission of an article for consideration for publication implies the transfer of the copyright from the author to the publisher upon acceptance.

Review Procedure

All submitted manuscripts are examined by the editorial staff for adequacy of documentation, composition and adherence to the guidelines of MPP. Manuscripts not submitted in accordance with these guidelines will be returned to the author for correction before beginning the review process. The manuscripts that are considered suitable for review are sent to at least two referees for evaluation. All reviews are conducted confidentially. The referees are asked to assess the originality, scientific merit, design of the study including statistical analysis, professional interest and the overall quality of the manuscript. The referee may recommend accept as is or with revision. It is unusual for a manuscript to be accepted without undergoing revision. Two copies of the revised manuscript are returned to the Editor-in-Chief for further processing. All accepted manuscripts are subject to editing for clarity, accuracy and style. All authors of accepted manuscripts are required to sign the Assignment of Copyright Agreement to consign the copyright of their paper to the publisher. Accepted manuscripts become the property of MPP and may not be reproduced by any means, in whole or in part, without the written consent of the publisher.

Arrangement

Title page: The first page of each paper should indicate the title, the authors' names, the institute where the work was conducted, and a short title for use as running head.
NB: Authors wishing to preserve the phonetic meaning of diacritics (PubMed reduces diacritics to their root characters) must spell their names accordingly when submitting manuscripts (e.g. Müller should be Mueller).

Full address: The exact postal address of the corresponding author complete with postal code must be given at the bottom of the title page. Please also supply phone and fax numbers, as well as e-mail address.

Key words: For indexing purposes, please list 3–10 key words.

Abstract: The abstract should be structured and not exceed 250 words. The abstract should include four clearly identifiable areas of content: *Objective* (the rationale and importance of the study), *Materials/Subjects and Methods* (description of the materials, animals or subjects, basic design, procedures and the analytical techniques), *Results* (major findings), *Conclusion* (significant implication of the finding).

Text: Manuscripts are typed double-spaced on one side of A4 paper, with 3-cm margins and limited to approximately 16 pages in length including references, tables and figures. Generic names should be used. MPP discourages the use of acronyms. The first time an abbreviated name or symbol is used, the full word or name of the item that is being abbreviated should precede it.
The text of an *original laboratory* or *clinical investigation* should comprise of Introduction and Objective, Materials and Methods (laboratory investigations) or Subjects and Methods (clinical investigations), Results, Discussion, Conclusion, References, Tables, Figure legends and Figures. Scientific measurements should be given in SI units. Any statistical procedure used should be either described in detail or supported by references.
The text of *Short Communications* should follow the above outline, and is limited to 5 manuscript pages (about 1,000 words) including references, tables and figures. Please provide a maximum of 7 references and 1 table or figure.
The text of *Case Reports* should comprise Introduction and Objective, Clinical Presentation including intervention or technique, Discussion including importance of the case reported, and Conclusion.
Reviews are generally solicited, should be of current interest and not more than 35 pages in length including references, tables and figures.
Letters should be comments on previously published articles or issues of interest to the public. They should be brief, concise and to the point. Letters commenting on previously published articles should be received within 1 year of the publication of the said article. All letters are subject to editing.

Acknowledgement: Please provide on a separate page if applicable.

Tables and illustrations: Tables and figures should be numbered using Arabic numerals. Prepare each table or figure on a separate page. Tables require a heading while figures require a legend. The legend should be prepared on a separate page. Only good drawings and original photographs will be accepted; negatives or photocopies will be rejected. Due to technical reasons, figures with a screen background should not be submitted. When possible, group several figures in one block for reproduction (max. size 180 × 223 mm) or provide crop marks. On the back of each figure, write its number, the author's name and 'top' with a soft pencil. Electronically submitted b/w half-tone and color illustrations must have a final resolution of 300 dpi after scaling, line drawings one of 800–1,200 dpi. Photomicrographs should be calibrated on the print by adding a calibration bar (e.g. – = 1 μm). Any labeling should be done on copies and not on the prints. Patients shown in photographs should have their identity concealed or should give their written consent to publication. If any tables or figures that are submitted have been published elsewhere, written consent for republication should be obtained by the author from the holders of the copyright (usually the authors and publishers).

Color illustrations: Up to 6 color illustrations per page can be integrated within the text at the special price of CHF/USD 660.00 per page. Color illustrations are reproduced at the author's expense.

References: In the text, identify references by Arabic numerals [in square brackets]. Manuscripts accepted for publication but not yet published may be cited but must be referred to as 'in press'. The name of the journal and year must be listed.
Number the references according to the order in which they are first mentioned in the text. Write the last names of the authors followed by initials. There should be no punctuation other than a comma to separate the authors. Preferably, please cite all authors. Abbreviate journal names according to the Index Medicus system. (Also see International Committee of Medical Journal Editors: Uniform requirements for manuscripts submitted to biomedical journals. N Engl J Med 1997;336:309–315. www.icmje.org)

Digital Object Identifier (DOI)
S. Karger Publishers supports DOIs as unique identifiers for articles. A DOI number will be printed on the title page of each article. DOIs can be useful in the future for identifying and citing articles published online without volume or issue information. More information can be found at www.doi.org.

Examples
(a) Papers published in periodicals: Chatel J-M, Bernard H, Orson FM: Isolation and characterization of two complete Ara h 2 isoforms cDNA. Int Arch Allergy Immunol 2003;131:14–18.
(b) Papers published only with DOI numbers:
Theoharides TC, Boucher W, Spear K: Serum interleukin-6 reflects disease severity and osteoporosis in mastocytosis patients. Int Arch Allergy Immunol DOI: 10.1159/000063858.
(c) Monographs: Matthews DE, Farewell VT: Using and Understanding Medical Statistics, ed 3, revised. Basel, Karger, 1996.
(d) Edited books: DuBois RN: Cyclooxygenase-2 and colorectal cancer; in Dannenberg AJ, Dubois RN (eds): COX-2. Prog Exp Tum Res. Basel, Karger, 2003, vol 37, pp 124–137.

Page Charges

There are no page charges for papers of 5 or fewer printed pages (including tables, illustrations and references). Each additional complete or partial page is charged to the author at CHF/USD 275.00. The allotted size of a paper is equal to approximately 16 manuscript pages (including tables, figures and references).

Proofs

Unless indicated otherwise, proofs are sent to the address given for correspondence and should be returned with the least possible delay. Alterations made in proofs, other than the correction of printer's errors, are not accepted.

Reprints

Order forms and a price list are sent with the proofs. Orders submitted after the issue is printed are subject to considerably higher prices.

KARGER

Fax +41 61 306 12 34
E-Mail karger@karger.ch
www.karger.com

The Guidelines for Authors are available at:
www.karger.com/mpp_Guidelines

General Information

Medical Principles and Practice

Publication data: 'Medical Principles and Practice'. Volume 14 with 6 issues appears in 2005.

Subscription rates: Subscriptions run for a full calendar year. Prices are given per year.

Personal subscription:

Print or Online	Print/Online combined
CHF 527.50	CHF 568.50
EUR 390.50	EUR 420.50
USD 479.50	USD 516.00
postage and handling	(added to print and print/online)
CHF 32.70 Europe,	CHF 48.60 Overseas
EUR 23.70	
USD 44.10	

Institutional subscription:

Print or Online	Print/Online combined
CHF 1055.–	CHF 1139.–
EUR 781.–	EUR 844.–
USD 959.00	USD 1035.00
postage and handling	(added to print and print/online)
CHF 40.80 Europe,	CHF 60.60 Overseas
EUR 29.40	
USD 55.20	

Airmail surcharge: CHF 40.80 / USD 33.90

Microforms: Please direct all enquiries and orders to:
University Microfilms Inc.
300 North Zeeb Road, P.O. Box 1346
Ann Arbor, MI 48106-1346 (USA)

Subscription orders: Orders can be placed at agencies, bookstores, directly with the Publisher
S. Karger AG, P.O. Box
CH–4009 Basel (Switzerland)
or with any of the following distributors:
Deutschland: S. Karger GmbH, Postfach
D–79095 Freiburg
France: Librairie Luginbühl
36, bd de Latour-Maubourg
F–75007 Paris
Great Britain, Ireland: S. Karger AG
4 Rickett Street
London SW6 1RU, UK
USA: S. Karger Publishers, Inc.
26 West Avon Road, P.O. Box 529
Farmington, CT 06085, USA
India, Bangladesh, Sri Lanka:
Panther Publishers Private Ltd.
33, First Main
Koramangala First Block
Bangalore 560 034, India
Japan: Karger Japan, Inc.
Yushima S Bld. 3F
4-2-3, Yushima, Bunkyo-ku
Tokyo 113-0034, Japan

Change of address: Both old and new address should be sent to the subscription source.

Single issues and back volumes: Information can be obtained through the Publisher.

Advertising: Correspondence and rate requests should be addressed to the Publisher.

Bibliographic indices: This journal is regularly listed in Index Medicus/MEDLINE, Science Citation Index, EMBASE/Excerpta Medica, IMEMR (WHO) and Reference Update.

Copying: This journal has been registered with the Copyright Clearance Center (CCC), as indicated by the code appearing on the first page of each article. For readers in the US, this code signals consent for copying of articles for personal or internal use, or for the personal or internal use of specific clients, provided that the stated fee is paid per copy directly to
Copyright Clearance Center Inc.
222 Rosewood Drive
Danvers, MA 01923 (USA)
A copy of the first page of the article must accompany payment. Consent does not extend to copying for general distribution, for promotion, for creating new works, or for resale. In these cases, specific written permission must be obtained from the copyright owner,
S. Karger AG,
P.O. Box, CH–4009 Basel (Switzerland).

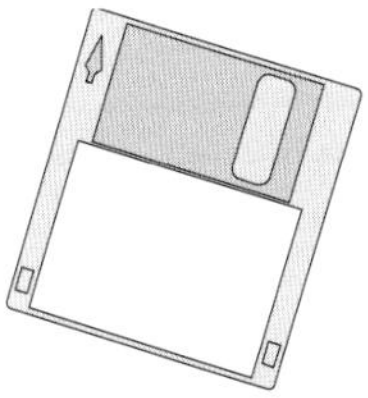

Disk Submission

A **comprehensive version** of these guidelines for the **submission of manuscripts and illustrations on disk** can be accessed online under **www.karger.com/disksubmission** or requested as a leaflet from the editorial department.

Please follow the general rules on style and arrangement in the **'Guidelines for Authors'** of the journal. Submit a **double-spaced printout** of your manuscript matching the file on disk exactly.

Technical Information
Store only data pertinent to this manuscript. Include your fonts (screen and printer fonts for Macintosh). For a **text only** manuscript, use a 3.5-inch disk. For **text and illustrations,** use 3.5 or 5.25 Magneto Optical Disc; CD-ROM (DOS or hybrid); iomega Zip/DOS (100 MB), or 44 MB, 88 MB or 200 MB SyQuest. Compress extensive data using ZIP format.

Data Description
Label your disk with the **title of the journal or book** for which it is submitted; a **short version** of the paper's title; your **name,** postal and e-mail address, fax and phone number; **hard- and software** used, including the version; and the **complete list of the file names.** File names consist of eight characters before the dot identifying the file and three characters after the dot identifying the application.

Text and Format
The **preferred word-processing package is MS-Word.** Other commonly used PC text programs and ASCII formats are accepted.
Update your data when you revise subsequent to the reviewer's or editor's decision and mark revisions clearly. Manuscripts accepted for publication are subject to **copy editing:** changes and/or questions will be marked in the manuscript, which will be returned to you with your disk and the galley proofs for checking.
Enter your text continuously flush left. Do not use any layout functions. Do not split words at the end of a line. **Do not use indentations.** Use a blank line before and a hard return at the end of paragraphs.
Arrange headings flush left. Insert only **one space after punctuation marks.** Insert Greek letters, mathematical symbols etc. with your word-processing program. Use **boldface and italics** as well as sub- and superscript where appropriate.
Tables are part of the text and should be placed at the end of the text file. Separate the individual parts of a table by lines. Use the tabulator instead of spaces. Format operation data on one line in the same column.

Legends to illustrations are part of the text and should also be listed at the end of the text file.

Illustrations
Store illustrations as separate files: do not integrate them in your text. Export **line drawings/vector graphics** as TIF, EPS or WMF format. Use Photoshop for processing and retouching **scanned half-tone images.** Save the original scan and the processed version. Export **black and white or color images** as TIF or EPS format in their anticipated size in print. **For detailed information** on how to submit illustration data on disk please consult the leaflet or **www.karger.com/disksubmission.**

KI 02 DISK

KARGER

Fax +41 61 306 12 34
E-Mail karger@karger.ch
www.karger.com

The Journal Home Page is available at:
www.karger.com/mpp

a b
HEALTHCARE
We don't just have a vision of the future.
We have the plan to get you there.
The Agfa-Gevaert Group is Europe's leading imaging company operating World Wide, developing, producing and distributing an extensive range of analogue and digital healthcare imaging systems, services and products.
AGFA offers a complete solution for radiology, 'from Analogue to Digital' including state of the art, films, processors, daylight and hardcopy systems, computed radiology (CR), radiology information systems (RIS) and picture archiving and communication systems (PACS).
شركة ياكو الطبية ش.م.ك.م.
YIACO Medical Co. K.S.C.C.
4842322 / 4814358 / 4821794 / 4832600 / 1 / 2 EXT. : 238 / 239 / 240 / 241
4844954 / 4833612 435, Safat 13005 - Kuwait
admin@yiacokuwait.com
inspire. transform. achieve
Agfa, Middle East Regional Office, P O Box 7143, Dubai, UAE, Tel: +97143965055, Fax: +97143965380.
F 05053

Medical Principles
and Practice

Proceedings of the International Conference on

Applications of Molecular Biology in Modern Medicine

Kuwait, March 7–11, 2004

Faculty of Medicine, Health Sciences Centre,
Kuwait University

Guest Editor
Christopher H.J. Ford, Safat, Kuwait

29 figures and 21 tables, 2005

KARGER

Basel · Freiburg · Paris · London · New York ·
Bangalore · Bangkok · Singapore · Tokyo · Sydney

S. Karger
Medical and Scientific Publishers
Basel • Freiburg • Paris • London
New York • Bangalore • Bangkok
Singapore • Tokyo • Sydney

Drug Dosage
The authors and the publisher have exerted every effort to ensure that drug selection and dosage set forth in this text are in accord with current recommendations and practice at the time of publication. However, in view of ongoing research, changes in government regulations, and the constant flow of information relating to drug therapy and drug reactions, the reader is urged to check the package insert for each drug for any change in indications and dosage and for added warnings and precautions. This is particularly important when the recommended agent is a new and/or infrequently employed drug.

P.O. Box, CH–4009 Basel (Switzerland)
Printed in Switzerland on acid-free paper by
Reinhardt Druck, Basel
ISBN 3–8055–8002–9

KARGER

Fax +41 61 306 12 34
E-Mail karger@karger.ch
www.karger.com

Vol. 14, Supplement 1, 2005

Medical Principles and Practice

Contents

KARGER

Fax +41 61 306 12 34
E-Mail karger@karger.ch
www.karger.com

Access to full text and tables of contents, including tentative ones for forthcoming issues: www.karger.com/mpp_issues

Medical Principles and Practice

Med Princ Pract 2005;14(suppl 1):1
DOI: 10.1159/000086178

Preface

A recent survey of biomedical publications from Arab countries showed that they produce less than 1% of the biomedical citations in the world, despite the available wealth and resources [1]. Small countries such as Kuwait, the United Arab Emirates and Lebanon were found to be more productive than larger countries, but biomedical research was found to focus mainly on curative services rather than research and development in areas such as molecular genetics. The ongoing revolution in the field of molecular biology makes it mandatory for all biomedical scientists to familiarize themselves with its implications for the practice of medicine. With this background, the Faculty of Medicine, Kuwait University, Kuwait, held an International Conference on the Applications of Molecular Biology in Modern Medicine, March 7–11, 2004. This supplement is based on the papers presented at the conference. The Central Organising Committee set the following goals for the conference:

1 to raise awareness about molecular biology and its applications with relevance to Kuwait in particular, and the Arabian Gulf in general;
2 to show to the outside world the research that has been done in Kuwait;
3 to provide training in some of the practical aspects of molecular biology;
4 to initiate and foster collaborative research between researchers in Kuwait and experts from other parts of the world.

The conference covered the applications of molecular biology in four areas – haematology, infectious diseases, cancer and metabolic diseases. High-calibre international, regional and local speakers were invited to give keynote addresses and there were also free oral presentations. The papers and abstracts presented in this supplement can be seen to have done full justice to the subject considering the breadth of presentations ranging from diagnosis, therapeutics, genotype/phenotype interaction to anthropology. It is noteworthy that the majority of the papers/abstracts came from investigators in Kuwait. On behalf of the Central Organising Committee, we should like to express our gratitude to all our sponsors especially: Kuwait University; Kuwait Foundation for the Advancement of Science (KFAS); Kuwait Institute for Medical Specializations (KIMS), and Kuwait Oil Corporation (KOC). We could not have produced this supplement without the generous financial support from the Company for Financial Center and Yousuf Ibrahim Al-Ghanim Company (YIACO), Kuwait. We are also indebted to the Editor-in-Chief, Professor Farida Al-Awadi, and the Editor, Professor Azu Owunwanne, for their help and unwavering commitment to the publication of this supplement. Last, but not least, we would like to thank all the participants at the conference and those who have generously given of their time to provide manuscripts. We hope that the content of these pages will go some way towards redressing the situation indicated by Tadmouri and Bissar-Tadmouri [1].

Prof. *Christopher H.J. Ford,* Chairman
Scientific Programme and Publication Committee

Dr. *Mohammad Z. Haider,* Chairman
Organising Committee

Reference

1 Tadmouri GO, Bissar-Tadmouri N: Biomedical publications in an unstable region: The Arab world, 1988–2002. Lancet 2003;362:1766.

KARGER

Fax +41 61 306 12 34
E-Mail karger@karger.ch
www.karger.com

1011–7571/05/0147–0001$22.00/0

Accessible online at:
www.karger.com/mpp

Review

Medical Principles and Practice

Med Princ Pract 2005;14(suppl 1):2–14
DOI: 10.1159/000086179

Received: August 15, 2004
Revised: October 25, 2004

Megaloblastosis: From Morphos to Molecules

K.C. Das[a] Monisha Das[b] D. Mohanty[b] M.M. Jadaon[a] A. Gupta[a] R. Marouf[a] S.K. Easow[a]

[a]Hematology Unit, Department of Pathology, Faculty of Medicine, Kuwait University, Kuwait;
[b]Institute of Immunohematology, Mumbai, India

Key Words
Megaloblastosis · Electron microscopy · Chromosomal despiralization · Autoradiography · ^{3}H-thymidine incorporation · DNA synthesis · Deoxyuridine suppression test · Histone biosynthesis

Abstract
Objective: Megaloblastosis (i.e., megaloblastic transformation of erythroid precursor cells in the bone marrow) is the cytomorphological hallmark of megaloblastic anemia resulting from vitamin B_{12} and folate deficiency. It is characterized by a finely stippled lacy pattern of nuclear chromatin, which is believed to be an expression of deranged cellular DNA synthesis. However, the molecular basis of these cytomorphological aberrations still remains obscure. The current presentation describes the results of our studies on some molecular events associated with the development of megaloblastosis. ***Methods:*** Transmission electron microscopy was used to study megaloblasts as well as DNA fibers extracted from megaloblastic and normoblastic bone marrows with and without treatment with proteinase K during the extraction procedure; cellular DNA synthesis in bone marrow cultures was studied by incorporation of ^{3}H-thymidine and deoxyuridine suppression test, while histone biosynthesis in bone marrow cells was studied by in vitro incorporation of ^{3}H-tryptophan, ^{3}H-lysine and ^{3}H-arginine into histones. ***Results:*** Derangement of DNA synthesis occurred due to an impaired de novo pathway of thymidylate synthesis in both vitamin-B_{12}- and folate-deficient human megaloblastic bone marrows as well as in the bone marrows of rhesus monkeys and rats with experimentally induced folate deficiency. Interestingly, folate-deficient monkeys developed frank megaloblastic bone marrows, but folate-deficient rats did not. On the other hand, megaloblastic changes in the bone marrow of human patients with myelodysplastic syndrome and erythroleukemia were not associated with this DNA synthetic abnormality. Biosynthesis of predominantly arginine-rich histones in megaloblastic bone marrows was markedly reduced as compared to normoblastic bone marrows, which was consistently associated with elongation and despiralization of chromosomes and finely stippled nuclear chromatin in megaloblasts. ***Conclusion:*** The impaired biosynthesis of predominantly arginine-rich nuclear histones appeared to be a common molecular event (a denominator) underlying the development of megaloblastosis with or without abnormal DNA synthesis.

Introduction

Megaloblastic anemia is essentially recognized as a morphological entity and is characterized by megaloblastic transformation of erythroid precursor cells in the bone

KARGER

Fax +41 61 306 12 34
E-Mail karger@karger.ch
www.karger.com

1011–7571/05/0147–0002$22.00/0

Accessible online at:
www.karger.com/mpp

Dr. K.C. Das
Hematology Unit, Department of Pathology, Faculty of Medicine, Kuwait University
PO Box 24923, 13110 Safat (Kuwait)
Tel. +965 531 9476, Fax +965 533 8905
E-Mail drkcdas@hotmail.com, kshitish@hsc.edu.kw

marrow of these patients. Megaloblastosis is the cytomorphological diagnostic marker of megaloblastic anemia, and represents a unique morphological and functional aberration of erythropoiesis. Megaloblasts are characterized by several cytomorphological features which are distinct from those of normal erythroid precursor cells (i.e., normoblasts) at all stages of maturity. These include larger cell size, and a finely stippled, open lacy pattern of nuclear chromatin. The cytoplasmic features are less remarkable, with evident asynchrony between nuclear and cytoplasmic maturation. The original prototype of this anemia was Addisonian pernicious anemia, initially recognized in the Caucasians as the only example of megaloblastic anemia in the nineteenth century. It was later discovered that this anemia was caused by a deficiency of vitamin B_{12} due to a lack of its obligatory binder gastric intrinsic factor and consequent malabsorption of this vitamin [1]. Lucy Wills [2, 3] described for the first time another type of megaloblastic anemia, the 'pernicious anemia of pregnancy' and 'tropical macrocytic anemia', which were later found to be due to a deficiency of folate (folic acid). It is, however, not infrequent to find strict vegetarians and vegans in the Indian subcontinent (either due to religious and social compulsions, food faddism or poverty) who develop vitamin B_{12} deficiency and megaloblastic anemia even when gastric intrinsic factor is not deficient [4, 5]. Common causes of megaloblastic anemias are deficiency of vitamin B_{12} and folate due to various reasons such as poor dietary intake, malabsorption, increased requirement, utilization and excretion [6]. Cytomorphological changes very similar to megaloblastosis can also occur in erythroid precursor cells in the bone marrow of patients with myelodysplastic syndrome and erythroleukemia; these latter conditions are relatively rare, and the nature of these 'megaloblastic' changes remains obscure. Whereas vitamin-B_{12}- and folate-deficient megaloblastic anemias are successfully treated by providing the deficient vitamins (i.e., folic acid or vitamin B_{12}), 'megaloblastic' changes associated with myelodysplastic syndrome and erythroleukemia are resistant to treatment with these vitamins. Megaloblastic changes caused by vitamin B_{12} and folate deficiency appear to be systemic processes not restricted to the cells of the erythroid series, but affecting all replicating cells [1, 7–9]. The appearance of giant metamyelocytes and hypersegmented neutrophils in vitamin-B_{12}- and folate-deficient megaloblastic anemias are also believed to be cytomorphological expressions of these vitamin deficiencies [1, 9].

When peripheral blood lymphocytes of patients with vitamin-B_{12}- or folate-deficient megaloblastic anemias are stimulated by a mitogen such as phytohemagglutinin in tissue cultures, they undergo 'blastic' transformation, but the cytomorphological features of these transformed lymphoblasts are different from similarly transformed lymphoblasts from normal subjects. The transformed lymphoblasts from megaloblastic anemia patients are generally larger, possess finely stippled nuclear chromatin, and also develop a DNA synthetic abnormality as revealed by an abnormal or impaired de novo pathway of thymidylate synthesis. They have been described as 'megaloblastic lymphocytes' [10, 11]. The cytomorphological changes in megaloblastic erythroid precursor cells and other replicating cell systems are generally believed to be morphological expressions of a poorly defined derangement of cellular DNA synthesis [7, 10–12]. This hypothesis has not since been adequately explored or verified in clinical or experimental conditions.

The current presentation is based on our studies on cytomorphological features and molecular events in human megaloblastic bone marrows, and in the bone marrows of two mammalian species (i.e., rhesus monkeys and Wistar strain rats) rendered folate deficient by dietary deprivation of folate. These studies included (a) cytomorphological studies of human megaloblastic bone marrows with particular focus on nuclear chromatin; (b) cytogenetic analysis for chromosomal pattern and replication; DNA content, DNA synthesis and deoxyuridine (dU) suppression test; (c) cytomorphological features, DNA synthesis and dU suppression test in the bone marrows of folate-deficient monkeys and rats, and (d) histone biosynthesis in megaloblastic and control (normoblastic) bone marrows.

Materials and Methods

Human Subjects

This study was carried out on 50 patients with megaloblastic anemia of moderate to severe degree (table 1). Among them, 40 had vitamin B_{12} deficiency (25 had pernicious anemia with positive Schilling's test, parietal cell antibodies and intrinsic factor antibodies, whereas 15 had nutritional deficiency of vitamin B_{12}, being vegetarian or vegans) and 10 had folate deficiency as indicated by subnormal serum and red cell folate levels either due to poor folate intake or malabsorption. There were 15 patients with myelodysplastic syndrome, who presented with persistent pancytopenia, and macrocytic red cells with a few hypolobulated and hypogranular neutrophils (pseudo-Pelger anomaly). Their bone marrows showed normocellularity or hypercellularity and varying degrees of dysplastic changes involving two or more cell lines; the erythroid precursor cells in their bone marrows showed significant, but varying degrees

Table 1. Basic hematological profile of 50 patients with megaloblastic anemia and 30 control subjects

	Mean cell volume, fl	Megaloblastic changes in the bone marrow	Type of vitamin deficiency	
			vitamin B_{12}	folate
Patients				
Severely anemic, Hb 20–70 g/l (n = 15)	110–118	++++	n = 10 35–80 ng/l	n = 5 serum folate (0.5–2.5 μg/l) red cell folate (50–86 μg/l)
Moderately anemic, Hb 71–96 g/l (25) (n = 35)	104–115	+++	n = 30 55–97 ng/l	n = 5 serum folate (1.0–3.0 μg/l) red cell folate (75.0–105 μg/l)
Control subjects				
Hb 140–160 g/l (n = 30)	78–86	none; all showed normoblastic erythropoiesis	290–850 ng/l	serum folate (8–20 μg/l) red cell folate (350–980 μg/l)

Table 2. Folate levels in serum, red cells and liver before and after feeding rats and rhesus monkeys on folate-poor diet

Experimental groups	Serum folate, μg/l		Red cell folate, μg/l		Liver folate, μg/g	
	initial	after 16 weeks	initial	after 16 weeks	initial	after 16 weeks
Control rats (n = 10)	16.3 ± 5.4	15.8 ± 6.5	552.4 ± 10.2	485.4 ± 15.5	4.5 ± 1.6	4.8 ± 2.1
Rats on folate-poor diet (n = 10)	17.2 ± 6.1	2.5 ± 0.5	470.5 ± 7.5	83.4 ± 0.7	4.7 ± 1.2	0.8 ± 0.3
Control monkeys (n = 5)	19.2 ± 5.2	18.5 ± 6.3	436.4 ± 10.1	455.4 ± 8.5	6.4 ± 2.2	6.0 ± 1.9
Monkeys on folate-poor diet (n = 5)	18.4 ± 6.4	2.5 ± 0.3	476.4 ± 9.5	76.5 ± 8.4	5.9 ± 1.7	1.2 ± 0.6

of megaloblastic changes. There were 5 patients with erythroleukemia (AML, M6), which were diagnosed by their peripheral blood count, bone marrow cytomorphology, cytochemical features and immunophenotyping characteristics; their bone marrows showed megaloblastic changes. Thirty subjects aged 10–40 years, who were hematologically normal, but had undergone bone marrow examination for unrelated disorders to exclude metabolic disorders (e.g. Gaucher's disease, Niemann-Pick's disease), or for chromosomal studies to exclude chromosomal mosaic, were studied as controls. Bone marrow aspiration and trephine biopsies were performed after obtaining informed consents. Twenty patients with moderate to severe iron deficiency anemia were studied for histone biosynthesis in the bone marrow and compared with those with megaloblastic anemia.

Experimental Folate Deficiency

Folate deficiency was produced in two mammalian species: (a) higher mammal: rhesus monkeys; (b) lower mammal: Wistar strain rats.

Folate Deficiency in Rhesus Monkeys. Five adult healthy rhesus monkeys *(Macaca mulatta)* each weighing 5–6 kg were caged individually in a well-lighted room, acclimatized to the laboratory conditions for 6–8 weeks. Folate deficiency was produced in these animals by feeding them a folate-deficient diet through a Ryle's tube as described by us previously [13]. Another 5 monkeys were separately caged and were fed a folate-rich diet (containing 100 μg folic acid per 100 g of the diet) to serve as the control group. The development of folate deficiency was monitored by measuring the levels of folate in the serum, red cells and finally in the liver. The monkeys fed a folate-poor diet showed consistent evidences of development of folate deficiency in 16–20 weeks, whereas the animals fed a folate-rich diet (the control group) showed no deficiency (table 2). At the end of the experiments, the monkeys were anesthetized, and in addition to collecting blood samples, bone marrows were aspirated from the posterior iliac spine for morphological studies, short-term tissue cultures, DNA synthesis, dU suppression test and other biochemical analyses.

Folate Deficiency in Rats. Folate deficiency was produced in 10 weanling (4 weeks old) Wistar strain rats weighing about 50–60 g each. Each rat was kept in a separate cage. Folate deficiency was produced by feeding them a folate-deficient diet as described previously [13, 14]. Another 10 rats of similar age, sex and weight were separately caged to serve as the control group. The control animals were pair-fed with an isocaloric, folate-rich diet containing 50 mg of folic acid per kilogram of the diet; the folate-deficient diet contained no measurable folic acid content. Preliminary experiments showed that the rats fed on this deficient diet developed folate deficiency after 12–16 weeks. This was indicated by consistently low levels of folate in the serum, red cells and the liver in these animals, whereas the control animals showed no significant decrease in folate levels (table 2). At the end of the experiments, the rats were anesthetized by administering anesthetic ether and sacrificed. The bone marrows were taken from the femur and processed for morphology, short-term tissue culture, DNA synthesis and dU suppression test and other biochemical analyses.

Hematological Studies

Routine hematological studies were performed by standard techniques [15]. Blood and bone marrow smears were stained by May-Grünwald-Giemsa stain for cell morphology.

Transmission Electron Microscopy of Human Bone Marrow Cells

Briefly, the following procedure was used. Aspirated bone marrow cells were collected in balanced salt solution, centrifuged to make a cell button, which was fixed in 3.0% glutaraldehyde solution for 2–3 h, washed in Millonig's buffer, pH 7.3, and then treated with 1.0% osmium tetroxide, washed again in Millonig's buffer for 3 h, appropriately dehydrated in increasing concentrations of ethanol and dried in absolute ethanol, then put in two changes of propylene oxide for 10 min each, and finally blocks were made in pure epoxy resin, from which ultra-thin sections were finally made for transmission electron microscopy in a JEOL 1200 EX.

DNA Synthesis and dU Suppression Test

A portion of the aspirated bone marrow was collected aseptically in cold heparinized 0.06 *M* Tris-buffered Hank's balanced salt solution, pH 7.4, and processed into a monolayered suspension for short-term culture; DNA synthesis was monitored by measuring incorporation of ^{3}H-thymidine (^{3}H-TdR) into DNA, and dU suppression test was done as described by us previously [16, 17]. The dU suppression values were calculated as follows:

$$\text{Percentage due suppression} = \frac{^3\text{H-TdR incorporation into DNA with dU}}{^3\text{H-TdR incorporation into DNA without dU}} \times 100$$

$$\text{dU suppression correction by the additions of folates or vitamin } B_{12} = \frac{^3\text{H-TdR incorporation into DNA with dU and folate and/or vitamin } B_{12}}{^3\text{H-TdR incorporation into DNA without dU}} \times 100$$

^{3}H-TdR incorporation into cellular DNA is a sensitive radionucleoside marker of DNA synthesis. The generation of deoxythymidine monophosphate (dTMP) and then deoxy-thymidine triphosphate (dTMP→dTTP) is of particular importance for this synthetic process due to relative specificity of the thymine base in the composition of DNA (fig. 8). The dU suppression test essentially measures the efficacy of the de novo pathway by studying the incorporation of ^{3}H-TdR into DNA (via the salvage pathway) in the presence of a standardized excess of dU added to the short-term tissue culture of bone marrow cells. Normally, the deoxyuridylate (dUMP) suppresses the incorporation of ^{3}H-TdR into cellular DNA to 10% of control cultures (i.e., cultures without added dU). In deficiency of folate and vitamin B_{12}, the de novo pathway is relatively inactive as a result of which the added dU fails to suppress the incorporation of ^{3}H-TdR into DNA (i.e., ^{3}H-TdR incorporation is higher than 10% of the control). The higher the dU suppression value, the more inactive or inefficient is the de novo pathway. This abnormality can be corrected to 10% or less of dU-suppressed values by the addition of the deficient vitamins (folate or vitamin B_{12}). While all folates including 5-methyl-tetrahydrofolate (THF) completely correct the dU suppression abnormality due to folate deficiency, a combination of vitamin B_{12} and 5-methyl-THF is required to correct the abnormality in vitamin B_{12} deficiency [6, 10, 15].

The incorporation of ^{3}H-TdR into individual cells was studied by autoradiography as described by us previously [18]. DNA content of individual cells in the bone marrow smears was measured after Feulgen staining in a Becton-Dickinson Cellular Imaging System (CAS 200). Cytogenetic analysis was done as previously described by us [19].

Extraction and isolation of DNA from bone marrow cells were done by the classical phenol-chloroform extraction method as described by Sambrook et al. [20]. Replicate samples from each bone marrow were subjected to treatment with proteinase K during the extraction procedure in order to remove the nuclear proteins such as histones from the DNA, whereas the other sample from each bone marrow was not treated with proteinase K. The DNA fibers were precipitated and dehydrated in ethanol and air-dried. These fibers were then fixed in 3.0% glutaraldehyde, washed in Millonig's buffer, dehydrated in ethanol and finally embedded in epoxy resin, and ultra-thin sections were made for transmission electron microscopy.

Histone Content and Histone Biosynthesis in Bone Marrow Cells

Histone content was measured and its biosynthesis studied by methods previously described [21–24]. The incorporations of ^{3}H-tryptophan, ^{3}H-lysine and ^{3}H-arginine into bone marrow cells in vitro were used as radioactive tracers for biosynthesis of histones.

Serum and Red Cell Folate and Vitamin B_{12}

Serum and red cell folate and vitamin B_{12} were estimated by radioisotope dilution techniques using appropriate binders for these two vitamins as described previously [25, 26].

Statistical Method

Statistical package for Social Sciences (SPSS Inc., Chicago, Ill., USA) was used for data processing; a p value ≤ 0.05 was used as cutoff level of significance. Student's t test was used for comparing means.

Results

Cytomorphological Changes in Blood and Bone Marrows

The patients with megaloblastic anemia due to deficiency of vitamin B_{12} or folate had variable degrees of anemia from moderate to severe, with or without thrombocytopenia or leukopenia (table 1). Macrocytosis and macro-ovalocytosis of red cells and hypersegmented neu-

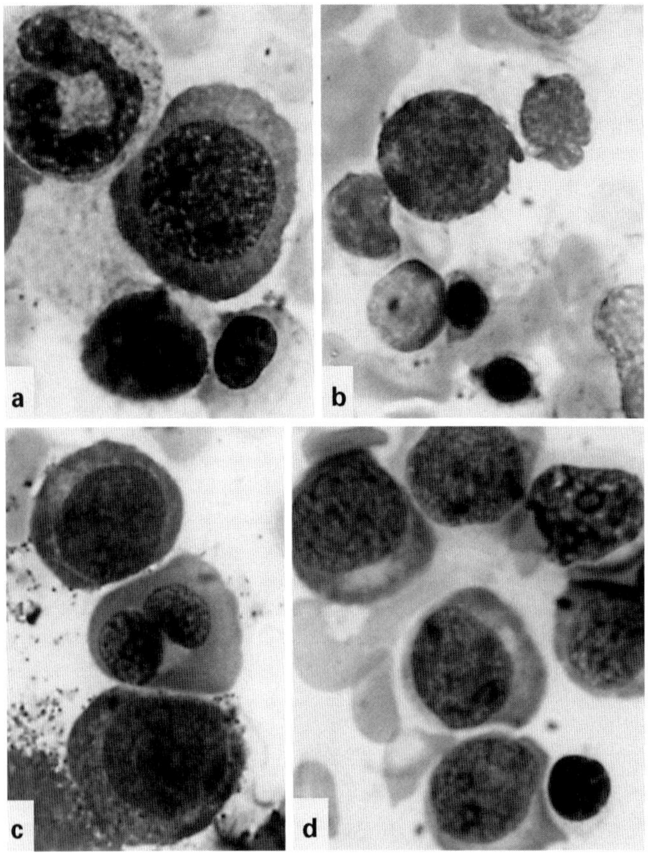

Fig. 1. Light microscopic cytomorphological features of megaloblasts (**a**) and normoblasts (**b**). Megaloblasts show relatively larger cell size, a finely stippled lacy pattern of nuclear chromatin, with asynchrony between nuclear and cytoplasmic maturation. In contrast, normoblasts show condensed and clumpy nuclear chromatin. Bone marrows from patients with myelodysplastic syndrome (**c**) and erythroleukemia (**d**) also show megaloblastic changes.

trophils were predominant morphological changes in peripheral blood. The bone marrow smears from the aspirates showed frank megaloblastosis of erythroid precursor cells, which were mainly characterized by larger cell size, and a finely stippled lacy pattern of nuclear chromatin, with relatively abundant cytoplasm and asynchrony between nuclear and cytoplasmic maturity (fig. 1a) as compared to normoblasts (from bone marrows of control subjects) (fig. 1b). The erythroid precursor cells in the bone marrow of patients with myelodysplastic syndrome and erythroleukemia also displayed variable degrees of immaturity of nuclear chromatin showing a finely stippled lacy pattern (fig. 1c, d), resembling the megaloblasts in vitamin-B_{12}- or folate-deficient megaloblastic anemia.

Table 3. Results of autoradiographic studies for incorporation of ^{3}H-TdR into DNA in bone marrow cells in vitro

	Labeling index	Mean grain count
Basophilic + polychromatic normoblasts	25 ± 6.7	21.57 ± 5.3
Basophilic + polychromatic megaloblasts	51.3 ± 8.2*	43.6 ± 8.7*

Control subjects: n = 30; vitamin-B_{12}-deficient subjects: n = 40; folate-deficient subjects: n = 10. * $p \leq 0.01$ compared to the normal values.

Transmission Electron Microscopy

These cytomorphological differences between the normoblasts and megaloblasts appeared magnified in the transmission electron microscopy with particular reference to the spongy, finely stippled appearance of the nuclear chromatin in the megaloblasts in contrast to condensed chromatin in the nuclei of normoblasts (fig. 2a–d).

Cytogenetic Studies

No significant numerical changes were observed; the model number of chromosomes was 46 and a very small number of metaphase plates showed random aneuploidy. However, the individual chromosomes showed striking structural changes, which were characterized by elongation and despiralization of chromosomes (fig. 3a, b); occasional polyploidy and endoreduplication were encountered. The replication pattern of chromosomes in megaloblastic anemia also appeared different from normoblastic bone marrows as studied by autoradiography after incorporation of ^{3}H-TdR into DNA (fig. 4a, b).

Cellular DNA Content, and DNA Synthesis (i.e., ^{3}H-TdR Incorporation into DNA)

The measurement of DNA content of megaloblasts and normoblasts by image analysis after Feulgen staining showed that the interphase (Go/G1) cells constituted the majority with modal DNA content of diploid cells (fig. 2c). A relatively small proportion of the cell population (<10%) showed random aneuploidy on either side of the diploid mode. Autoradiographic studies after 1 h of pulse exposure to ^{3}H-TdR in short-term culture of bone marrow cells showed that the labeling index as well as grain count of basophilic and polychromatic erythroblasts were significantly higher in the megaloblastic bone marrows than in the normal bone marrows (table 3); the orthochromatic

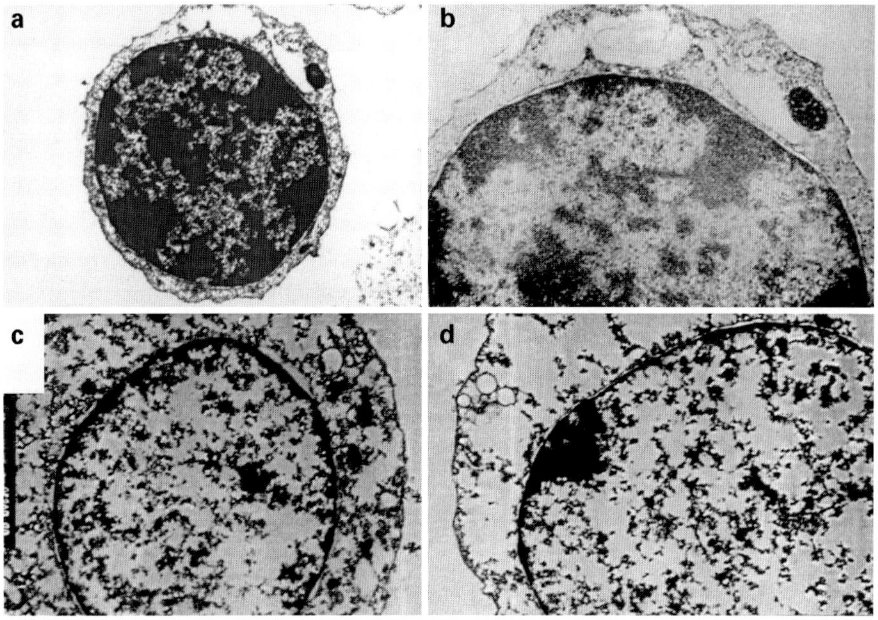

Fig. 2. Transmission electron microscopic picture of normoblasts (**a**, **b**) shows condensed and clumpy chromatin; in contrast, transmission electron microscopic picture of megaloblasts (**c**, **d**) shows well-dispersed finely stippled rarefied nuclear chromatin.

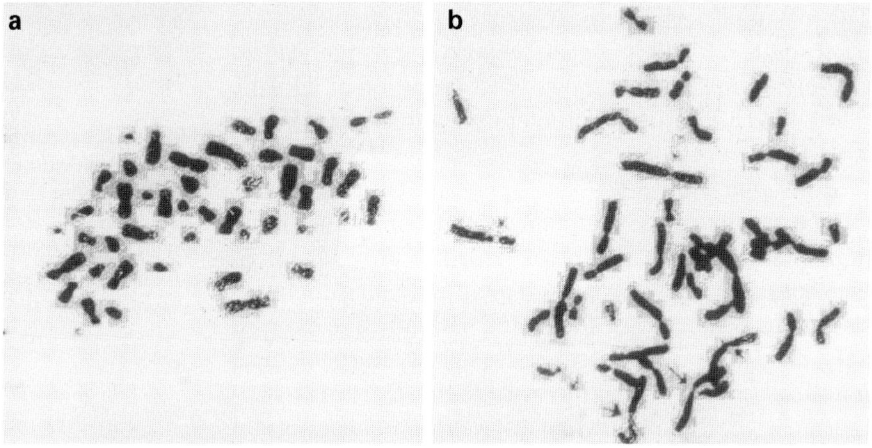

Fig. 3. A metaphase plate from bone marrow of a control (normal) subject (**a**) and from a patient with megaloblastic anemia (**b**) showing elongation and despiralization of individual chromosomes.

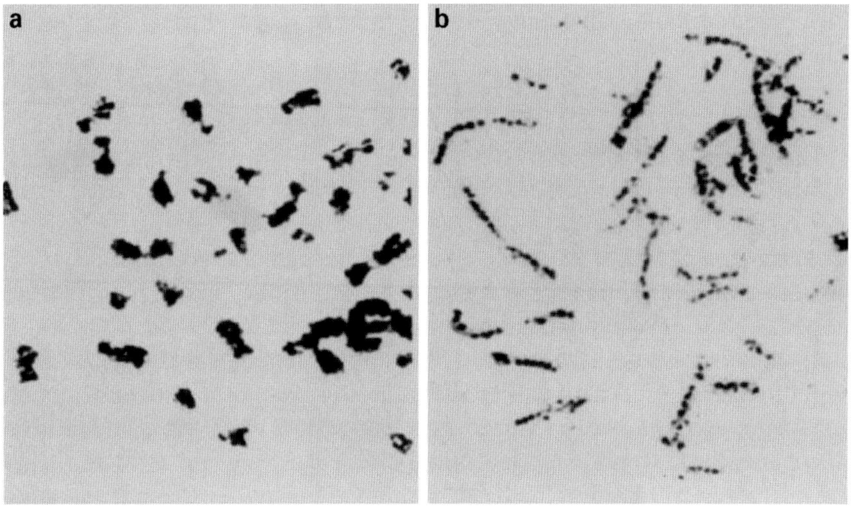

Fig. 4. Autoradiographic study of chromosomal preparations after ^{3}H-TdR incorporation showing the distribution pattern of silver grains due to ^{3}H-TdR uptake in normal (**a**) and megaloblastic chromosomes (**b**).

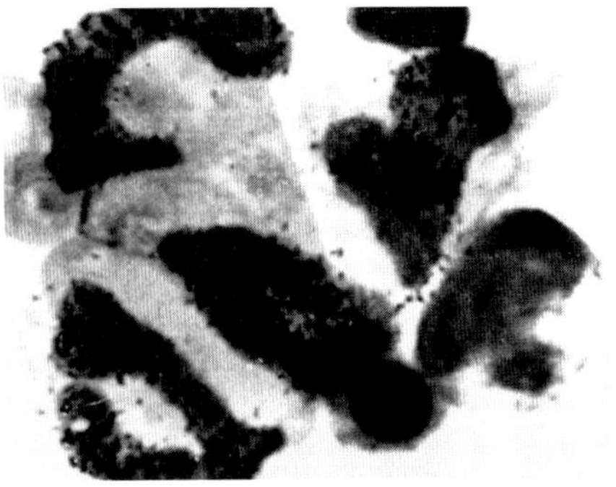

Fig. 5. Autoradiography of megaloblastic bone marrows showing ^{3}H-TdR-labeled giant metamyelocytes indicating their proliferative ability.

erythroblasts did not take up ^{3}H-TdR as these cells do not proliferate. The cells of granulocytic series in megaloblastic anemia also showed relative slowing of maturation. The metamyelocytes in the control bone marrows did not incorporate ^{3}H-TdR, but giant metamyelocyte in the megaloblastic bone marrow showed incorporation of ^{3}H-TdR into DNA, indicating its proliferative ability (fig. 5).

dU Suppression Test

The results of dU suppression tests in human bone marrows are shown in table 4. This has proved to be a very sensitive test to detect and quantitate megaloblastosis due to deficiency of folate and vitamin B_{12}, even in very mild or early clinical states when morphological manifestations remain equivocal [7, 10–12, 16, 17]. Furthermore, the dU suppression test was found to be normal when performed with bone marrows of patients with myelodysplastic syndrome and erythroleukemia (table 4) in which cytomorphological changes very similar to megaloblastosis also occurred.

Morphology and DNA Synthesis in the Bone Marrows of Folate-Deficient Animals

Folate-Deficient Rhesus Monkeys. Five adult monkeys fed a folate-poor diet for 4–5 months developed hematological and biochemical evidences of folate deficiency. They developed mild to moderate anemia and thrombocytopenia with or without leukopenia. The peripheral blood smear showed macrocytic red cells and hypersegmented neutrophils. Serum and red cell folate as well as liver folate were markedly reduced as compared to the control monkeys, which were fed a folate-rich diet (table 2). The serum vitamin B_{12} levels remained normal.

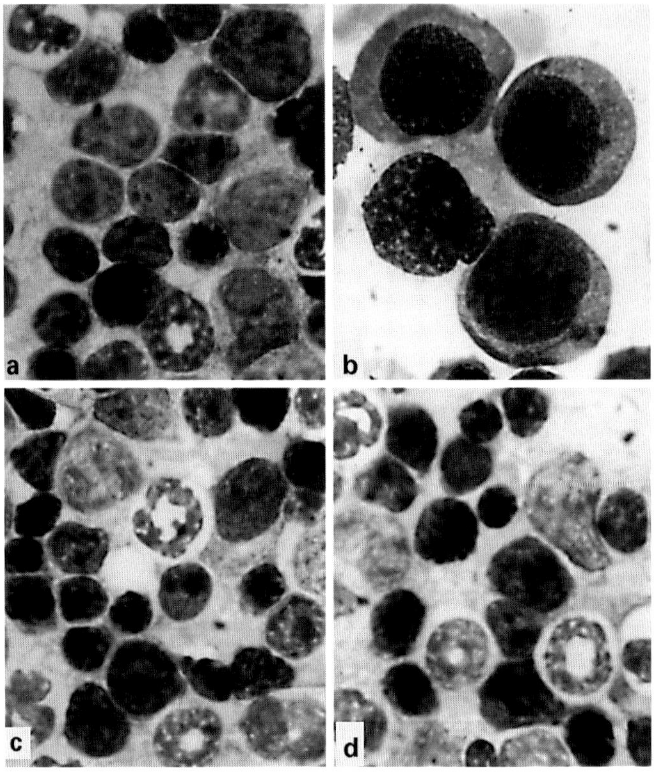

Fig. 6. a A bone marrow smear from a control rhesus monkey showing normoblastic erythropoiesis. **b** A bone marrow smear from a folate-deficient rhesus monkey showing florid megaloblastic transformation. **c** A bone marrow smear from a control rat showing normoblastic erythropoiesis. **d** A bone marrow smear from an established folate-deficient rat showing normoblastic erythropoiesis. Absence of megaloblastic transformation.

The bone marrow smears of the control monkeys fed a folate-rich diet showed normoblastic erythropoiesis (fig. 6a), whereas the folate-deficient monkeys revealed frank megaloblastic transformation (fig. 6b). The dU suppression test of their aspirated bone marrows showed abnormal results, which were corrected in vitro by the addition of 5-methyl-THF but not by vitamin B_{12} (table 5). All of these features were similar to those of human megaloblastic anemia due to folate deficiency.

Folate-Deficient Rats. Weanling Wistar rats fed a folate-poor diet developed persistently low levels of folate in their serum, red cells and liver in 12–16 weeks (table 2). They also developed mild to moderate anemia, thrombocytopenia and leukopenia, but red cells were still normocytic and normochromic. The bone marrow smears of control rats (fig. 6c) as well as of folate-deficient rats (fig. 6d) showed normoblastic erythropoiesis (i.e., ab-

Table 4. Summary of results of dU suppression tests in short-term cultures of bone marrow cells

	dU alone	dU + 5-methyl-THF	dU + CNC bl	dU + CNC bl + 5-methyl-THF	dU + folic acid
Normal (n = 30)	7.5 ± 2.2	6.8 ± 1.8	7.1 ± 2.0	8.5 ± 2.3	9.0 ± 1.4
Vitamin B_{12} deficient* (n = 40)	75.5 ± 15.0	59.6 ± 12.7	46.5 ± 15.2	9.2 ± 3.2	8.6 ± 2.4
Folate deficient* (n = 10)	63.8 ± 18.5	8.9 ± 3.2	61.9 ± 21.0	7.5 ± 1.8	9.0 ± 2.1
Myelodysplastic syndrome (n = 15)	7.2 ± 1.8	9.3 ± 2.4	6.9 ± 2.1	7.7 ± 2.5	8.2 ± 2.0
Erythroleukemia (n = 5)	9.5 ± 2.6	6.8 ± 1.9	7.5 ± 2.3	8.2 ± 2.2	7.4 ± 1.6

dU suppression values expressed as percentage incorporation of ^{3}H-TdR into DNA without dU (see sections on Methods). CNC bl = cyanocobalamin.

* $p < 0.01$ compared with normal values. The values for myelodysplastic syndrome and erythroleukemia were not significant compared to the normal values. Results are means ± SD.

Table 5. Results of dU suppression test in folate-deficient rats and rhesus monkeys

	dU alone	dU + folic acid	dU + 5-methyl-THF	dU + vitamin B_{12}
Control rats (n = 10)	10.2 ± 8.2	9.2 ± 2.6	9.5 ± 2.2	8.6 ± 2.8
Folate-deficient rats (n = 10)	38.5 ± 8.5*	9.8 ± 3.0*	8.9 ± 2.8*	40.5 ± 12.5*
Control monkeys (n = 5)	9.1 ± 2.2	10.1 ± 2.5	8.6 ± 1.8	7.9 ± 3.3
Folate-deficient monkeys (n = 5)	52.6 ± 15.6*	9.8 ± 3.5*	10.0 ± 2.9*	49.5 ± 11.8*

dU suppression values are expressed as percentage of ^{3}H-TdR incorporation into DNA (^{3}H-TdR incorporation in replicate cultures without dU is taken as 100%).

* $p < 0.01$; folate-deficient rats compared with control rats and folate-deficient monkeys compared with control monkeys.

sence of megaloblastic transformation). DNA synthesis as indicated by dU suppression test was abnormal or impaired, which was corrected by the addition of methyl-THF, not by vitamin B_{12} (table 5).

Transmission Electron Microscopy of DNA Fibers from Human Patients with Megaloblastic Anemia and from Control Subjects

DNA fibers from normoblastic bone marrows (i.e., bone marrows of control subjects) after extraction without proteinase K appeared tightly knotted or beaded at regular intervals yielding a wavy pattern (fig. 7a). When DNA was extracted from these cells with proteinase K, which removed nuclear proteins including histone, the DNA fibers appeared straight, elongated with disappearance of the knots and the wavy pattern (fig. 7b). On the other hand, DNA fibers extracted from megaloblastic bone marrow cells without proteinase K showed very few or no beads and knots and consequently looked relatively straightened and relaxed (fig. 7c) and comparable to the DNA fibers extracted from these cells using proteinase K (fig. 7d), suggesting that megaloblasts had a markedly reduced nuclear histone content.

Histone Biosynthesis in the Bone Marrows of Control Subjects and Patients

The incorporation of the radioactive amino acid ^{3}H-tryptophan is a marker of total histone biosynthesis, whereas the incorporation of ^{3}H-lysine occurs preferentially into lysine-rich histones (i.e., H_1, H_2A, and H_2B histones), and that of ^{3}H-arginine into arginine-rich histones (i.e., H_3 and H_4 histones). The results of these studies are shown in table 6. There was a decrease in total histone synthesis in the folate- and vitamin-B_{12}-deficient megaloblastic bone marrows as compared to those of controls and iron deficiency anemia. This defect appeared to have predominantly affected the synthesis of arginine-rich histones. Similar impairment of histone biosynthe-

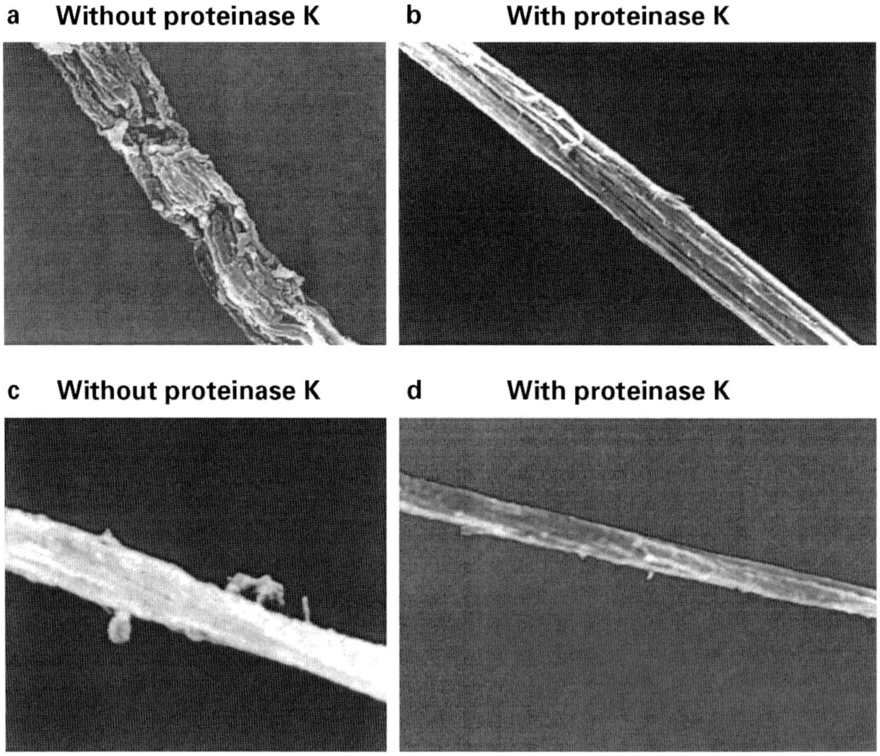

Fig. 7. Transmission electron microscopic picture of DNA fibers from a control (normal) bone marrow without proteinase K treatment showing knotted and spiral wavy pattern (**a**); DNA fiber from the control (normal) bone marrow after extraction with proteinase K to remove nuclear proteins (e.g. histones) showing that DNA fibers lost their knotted, beaded appearance and appeared relaxed and straight (**b**), from the bone marrow of a patient with megaloblastic anemia without proteinase K treatment during extraction (**c**), and with proteinase K treatment (**d**); both **c** and **d** showed loss of knots and wavy pattern resulting in despiralized and relaxed appearance of the fibers, suggesting absence or reduction of nuclear histones in the DNA of megaloblasts.

Table 6. The incorporation of ^{3}H-tryptophan, ^{3}H-lysine and ^{3}H-arginine into nuclear histones

	^{3}H-tryptophan	^{3}H-lysine	^{3}H-arginine
Control subjects (n = 20)	21,500 ± 1,510	11,448 ± 1,200	10,175 ± 1,125
Iron deficiency anemia (n = 20)	22,054 ± 1,325	10,955 ± 1,175	10,115 ± 1,096
Vitamin-B_{12}-deficient megaloblastic anemia (n = 30)	16,455 ± 1,024*	7,995 ± 875*	2,785 ± 615**
Folate-deficient megaloblastic anemia (n = 10)	15,145 ± 1,086*	8,155 ± 920*	3,225 ± 865**
Erythroleukemia (n = 5)	17,508 ± 512*	9,526 ± 726*	5,868 ± 497*
Myelodysplastic syndrome (n = 15)	16,973 ± 995*	8,972 ± 698*	6,287 ± 586*

The figures indicate the radioactivity (disintegrations per minute) incorporated. Results are means ± SD.
* $p < 0.05$; ** $p < 0.01$, compared to control subjects.

sis, particularly of arginine-rich histones, was also observed in the bone marrow of patients with myelodysplastic syndrome and erythroleukemia.

Discussion

Several hematological conditions are known to be associated with megaloblastic changes in the bone marrow. The most common among these are deficiency of vitamin B_{12} and folate, resulting in megaloblastic anemia, which can be completely corrected or cured by treating them with the deficient vitamins. Megaloblastic changes also occur in the bone marrow of patients with myelodysplastic syndrome and erythroleukemia, which are preneoplastic and neoplastic conditions, respectively; megaloblastosis in these two latter conditions are not reversed by therapy with vitamin B_{12} and folate [1, 7, 9]. Irrespective of their causes, megaloblastosis in all of these conditions as stated above are cytomorphologically similar, character-

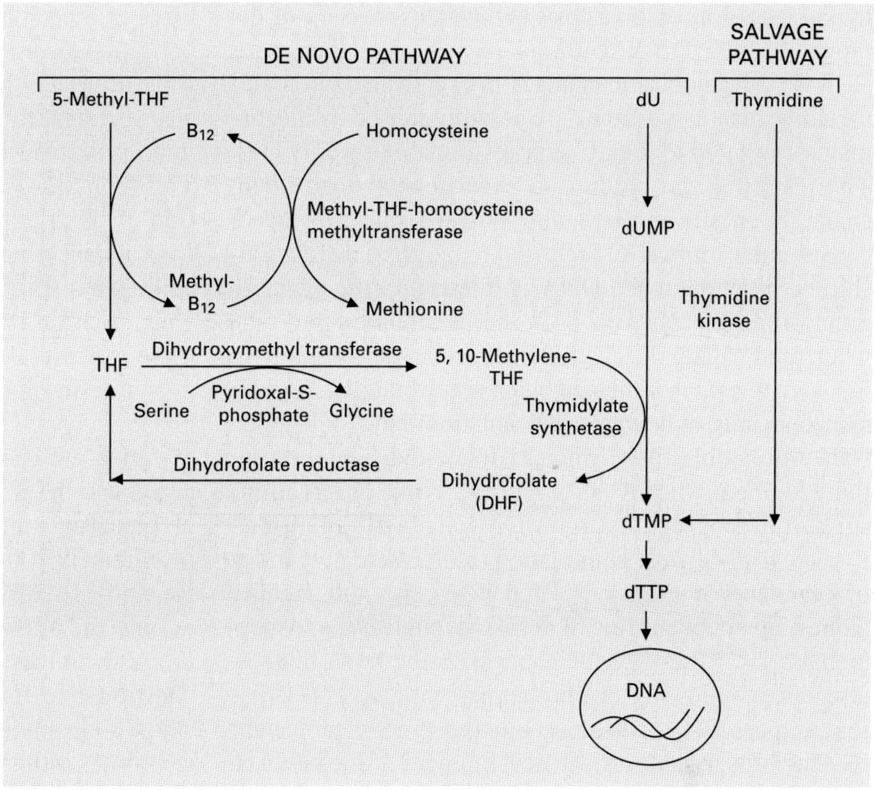

Fig. 8. A cartoon showing metabolic interrelations between vitamin B_{12} and folate with special reference to the de novo and salvage pathways of thymidylate synthesis (dUMP→dTMP).

ized by the open lacy pattern and spongy nuclear chromatin in both light and transmission electron microscopic observations. The distinctive structural changes in the chromatin of megaloblasts in patients with vitamin B_{12} and folate deficiency were found to be associated with several structural abnormalities of individual chromosomes such as elongation and despiralization (unwinding) of chromosomes. These may cause a defect in packaging chromosomes inside the cells, possibly impairing cell growth and cell functions, although no significant abnormalities of chromosome number were observed and the modal chromosome number remained 46 (diploid) with no consistent or significant aneuploidy [19, 27–29]. It has been suggested that the cytomorphological and cytogenetic changes in megaloblastic anemia may be an expression of deranged cellular DNA synthesis [7–12].

Cellular DNA comprises four deoxynucleotide triphosphate building blocks, namely deoxyadenosine triphosphate, deoxyguanosine triphosphate, deoxycytosine triphosphate and dTTP. In the double-stranded DNA helix, adenine pairs with thymine (A=T) and guanine with cytosine (G=A). The base composition of DNA extracted from megaloblastic and normoblastic bone marrows has been shown to be similar and the ratios of A:T and G:C are close to unity [30]. The synthesis of DNA in replicating cells is regulated by a number of reaction loops some of which are directly or indirectly dependent on the availability of folate coenzymes and vitamin B_{12} (fig. 8). The relevant and important reactions include: homocysteine to methionine interconversion, which requires methylcobalamin (CH_3-B_{12}) as an essential cofactor; in this reaction, 5-methyl-THF, the predominant form of plasma folate which is metabolically inactive, is also converted into an active compound, THF. This latter compound is, in turn, converted into 5,10-methylene-THF, which is an essential coenzyme in converting dUMP to dTMP by the enzyme thymidylate synthase. The dU suppression test provides a sensitive quantitative assessment of megaloblastic anemia and can also reveal the type of vitamin deficiency causing this anemia. This test is capable of distinguishing megaloblastosis associated with myelodysplastic syndrome and erythroleukemia from that resulting from deficiency of vitamin B_{12} and folate, since the dU suppression test is normal in the former two conditions, whereas it is abnormal in deficiency of vitamin B_{12} and folate. These observations also militate against the hypothesis of megaloblastosis being an expression of deranged cellular DNA synthesis, regardless of the causes [1, 7–9]. Megaloblastic changes in erythroleukemia and myelodysplastic syndrome are possibly caused by mechanisms unrelated to the dU suppression abnormality [18]. The results of our studies on experimental folate deficiency in rhesus monkeys and Wistar strain rats are of considerable interest and relevant to the possible relationship between deranged DNA synthesis and the development of megaloblastosis. Both rhesus monkeys and rats, when made folate deficient, developed an abnormality in DNA synthesis as indicated by an abnormal dU suppression test, but only folate-deficient monkeys developed frank megaloblastosis, whereas folate-deficient rats remained unequivocally normoblastic. These observations support the concept that while DNA synthetic abnormality as revealed by dU suppression test is a sensitive index of folate and/or vitamin B_{12} deficiency, this abnormality alone may not be the sole molecular event leading to megaloblastosis. Megaloblastic changes in the bone marrow also occur following chemotherapy with several drugs that impair DNA synthesis; but the precise molecular mechanisms underlying the development of megaloblastosis in those circumstances have not been adequately investigated.

The elongation and despiralization of chromosomes in replicating megaloblasts [19, 27–29] appear to reflect their characteristic cytomorphological features in the interphase nuclear chromatin as observed in light and transmission electron microscopy. The results of our studies on electron microscopic configuration of DNA fibers extracted from megaloblastic and normoblastic bone marrows appear to provide additional data on molecular events associated with megaloblastosis. The extraction of histones from DNA of normal bone marrow cells by treatment with proteinase K during extraction procedure made the DNA fibers lose the knotted and spiral pattern, and appear straight with bundles of fibers running parallel as observed by transmission electron microscopy; on the other hand, DNA fibers extracted from megaloblastic bone marrows, even when untreated by proteinase K, appeared relatively straight and devoid of knots and without wavy or spiral pattern. These findings suggest that the DNA extracted from megaloblastic bone marrows even without proteinase K treatment had a markedly decreased histone content as compared to DNA from normoblastic bone marrows (fig. 7a–d). These assumptions were further supported by our studies on histone biosynthesis. The results of incorporation of ^{3}H-tryptophan, ^{3}H-lysine and ^{3}H-arginine into nuclear histone in short-term cultures of megaloblastic as well as control (normoblastic) bone marrows provided a differential profile (table 6). Biosynthesis of histones in megaloblastic bone marrows was significantly lower than in control bone marrows or in bone marrows of patients with iron deficiency anemia. This reduction was more marked in respect of incorporation of ^{3}H-arginine than in respect of ^{3}H-lysine and ^{3}H-tryptophan. Nuclear histones play a significant role in coiling and spiralization of chromosomes, and considerable attention has generally been focused on the role of these proteins in the regulation of the structure of chromosomes in replicating cells and chromatin in the interphase nucleus. There are several species of histones, some of which are lysine rich (H_1, H_2A, H_2B) and others are arginine rich (H_3, H_4). The nuclear chromatin is made up of repeating units (nucleosomes), each containing approximately 200 bp of DNA and two each of H_2A, H_2B, H_3 and H_4. Most of the DNA molecules are winded around the outside of a core of histones – the remaining DNA, the linker DNA, joins adjacent nucleosomes – and contribute to the flexibility of the chromatin network. Thus, a chromatin is delicately joined to a chain of nucleosomes rather like beads on a string [21–23]. Changes in the composition of histones have been described in developing sperms [22] and duck erythroblasts [24], and in both instances, arginine-rich histones have been shown to increase as a function of cell maturation. The decreased synthesis of predominantly arginine-rich histones in the megaloblasts may be related to their impaired nuclear maturation and asynchrony of nuclear cytoplasmic development. It may be suggested that cytomorphological changes of megaloblastosis may be closely related to an imbalance in the molecular species of histones, which in some instances (as in vitamin B_{12} and folate deficiency) may interact with slowly synthesizing DNA in the replicating cells. The reduced synthesis of particularly arginine-rich histone may be a common denominator of biochemical events in clinical and hematological conditions associated with megaloblastic transformation of erythropoiesis. In recent years, several groups of investigators observed significant associations of specific histone patterns, acetylase/deacetylase activities with cell cycle changes, and apoptosis in many neoplastic cell models [31, 32]. Intranuclear mechanisms that signal apoptosis after DNA damage overlap with those that initiate cell cycle arrest and disturb cell cycle checkpoints [32, 33]. Interactions between these nuclear events and apoptotic processes may cause genomic instability in other cell systems as well [33, 34]. Naturally occurring or synthetic inhibitors of histone deacetylases can cause cell cycle ar-

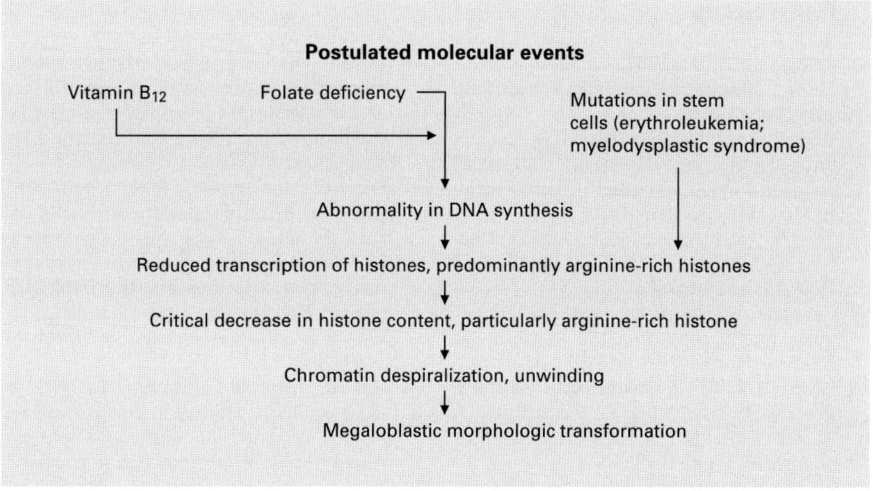

Fig. 9. A cartoon showing the postulated molecular basis of development of megaloblastosis.

rest, differentiation and apoptosis in neoplastic cells opening a door to explore newer cancer chemotherapy strategies [34, 35]. It is possible that changes in nuclear histones coupled with altered chromatin may also impair regulation of cell cycle checkpoints promoting apoptosis in megaloblastic bone marrows as suggested by several recent studies [36–39]. Recently, it has been reported that the nuclear histone content of bone marrows in lower mammalian species such as rats is much higher than that of higher mammalian species such as rhesus monkey [40]. This is believed to be a phylogenetic characteristic. When folate deficiency was induced in monkeys, their bone marrow developed florid megaloblastic transformation, and this was associated with marked reduction of particularly arginine-rich histones, but folate-deficient rats did not develop megaloblastic transformation, although DNA synthesis was impaired as indicated by abnormal dU suppression values. This can be explained by the initially high histone content in the bone marrow cells in the rats, which was not sufficiently reduced even in established folate deficiency to reach a critically low level required for megaloblastic transformation [36]. A postulated scheme of molecular events in the development of megaloblastosis is given in figure 9.

Conclusions

Cytomorphological features of megaloblasts in light and electron microscopy and the chromosomal pattern suggest that elongation, despiralization and uncoiling of chromosomes were related to the finely stippled lacy pattern of the nuclear chromatin which distinguishes them from normoblasts. DNA synthetic abnormality as revealed by abnormal dU suppression test in vitamin-B_{12}- and folate-deficient megaloblastic anemia is not a consistent molecular event underlying the development of megaloblastosis in all clinical and hematological disorders in which megaloblastic changes are observed (e.g. myelodysplastic syndrome and erythroleukemia). Nuclear proteins, particularly histones, are known to play a significant role in causing spiralization, twisting or coiling of chromosomes in the replicating cells and in the condensation of nuclear chromatin in the interphase. The reduced biosynthesis of histones particularly of arginine-rich histones as shown in 'megaloblastic' bone marrows irrespective of their causes appears to be a common denominator of molecular events associated with the development of megaloblastosis.

Acknowledgements

This work was supported by Kuwait University Research Grant No. MG 011 and the Institute of Immunohematology, Mumbai, India. The authors highly appreciate the excellent work done and the painstaking care taken by Mr. James Luke, Secretary, Department of Pathology, in preparing the manuscript.

References

1 Chanarin I: The Megaloblastic Anaemia, ed 3. Oxford, Blackwell Scientific Publications, 1990, pp 315–331.
2 Wills L: Treatment of pernicious anaemia of pregnancy and 'tropical anaemia' with special reference to yeast extract as a curative agent. Br Med J 1931;1:1059–1064.
3 Wills L: Treatment of pernicious anaemia, nutritional macrocytic anaemia and tropical sprue. Blood 1948;3:36–56.
4 Chatterjea JB: Nutritional megaloblastic anaemia in tropical zones. Proc 11th Congr Int Soc Hematol, Sidney, 1966, pp 120–132.
5 Das KC: Nutritional megaloblastic anaemia. Proc Symp Nutr Anemias, Plenary Sess 2, Meet Asian Pacific Div, Int Soc Hematol, Melbourne, 1971, pp 11–13.
6 Herbert V: The five possible causes of all nutrient deficiency. Aust NZ J Med 1972;1:69–72.
7 Das KC, Herbert V: Vitamin B_{12}-folate interrelations. Clin Haematol 1976;5:997–725.
8 Herbert V: The Megaloblastic Anemias, ed 1. New York, Grune and Stratton, 1959, p 1.
9 Lee RG: Megaloblastic anemias: Disorders of impaired DNA synthesis; in Lee RG, Foester J, Lukens J, Parakevas JP, Rodgers GM (eds): Wintrobe's Clinical Hematology, ed 10. Baltimore, Williams & Wilkins, 1999, pp 941–978.
10 Das KC, Hoffbrand AV: Lymphocyte transformation in megaloblastic anaemias. Morphology and DNA synthesis. Br J Haematol 1970; 19:459–468.
11 Das KC, Herbert V: The lymphocyte as a marker of past nutritional status. Persistence of abnormal lymphocyte deoxyuridine (dU) suppression test and chromosomes in patients with past deficiency of folate and vitamin B_{12}. Br J Haematol 1978;38:219–223.
12 Metz J, Kelly A, Swett VC, Waxman S, Herbert V: Deranged DNA synthesis by bone marrow from vitamin B_{12}-deficient humans. Br J Haematol 1968;14:575–592.
13 Mohanty D, Das KC: Effect of folate deficiency on the reproductive organs of female Rhesus monkeys: A cytomorphological and cytokinetic study. J Nutr 1982;112:1505–1576.
14 Johnson EM, Nelson MM, Monie IN: The composition of folate-deficient diet for rats. Anat Rec 1963;146:215.
15 Dacie JV, Lewis SM: Practical Haematology, ed 8. London, Churchill Livingstone, 1995, pp 49–143.
16 Das KC, Herbert V: In vitro DNA synthesis by megaloblastic bone marrow: Effect of folates and cobalamins on thymidine incorporation and de novo thymidylate synthesis. Am J Hematol 1989;31:11–20.
17 Das KC, Manusselis C, Herbert V: In vitro DNA synthesis by bone marrow cells and PHA-stimulated lymphocytes. Suppression by non radioactive thymidine of the incorporation of ^{3}H-deoxyuridine into DNA: Enhancement of incorporation when inadequate vitamin B_{12} and folate is corrected. Br J Haematol 1980;44:51–63.
18 Das KC, Garewal G, Mohanty D: Derangement of DNA synthesis in erythroleukemia. Normal deoxyuridine suppression and impaired thymidine incorporation in bone marrow culture. Acta Haematol 1980;64:121–134.
19 Das KC, Mohanty D, Garewal G: Cytogenetics in nutritional megaloblastic anaemia: Prolonged persistence of chromosomal abnormalities in lymphocytes after remission. Acta Haematol 1986;76:146–154.
20 Sambrook J, Fritsch EF, Maniatis J: Molecular Cloning: A Laboratory Manual. New York, Cold Spring Harbor Laboratory Press, 1989, pp 1–25.
21 Murray K: The acid extraction of histones from calf thymus deoxyribonucleoprotein. J Mol Biol 1966;15:409–419.
22 Das KC, Kaufman RP, Gay H: Autoradiographic evidence of synthesis of an arginine-rich histone during spermatogenesis in *Drosophila melanogaster*. Nature 1964;204: 1008–1009.
23 Tidwell T, Allfrey VF, Mirsky AE: The methylation of histone during regeneration of the liver. J Biol Chem 1968;243:707–715.
24 Gershey FL, Halett GW, Vidali G, Allfrey VG: Chemical studies on histone methylation. Evidence for the occurrence of 3-methyl-histidine in avian erythrocyte histone fraction. J Biol Chem 1969;244:4871–4877.
25 Longo DL, Herbert V: Radioassay for serum and red cell folate. J Lab Clin Med 1976;87: 138–151.
26 Das KC, Manusselis, Herbert V: Determination of vitamin B_{12} (cobalamin) in serum and erythrocytes by radioassay and of holo-transcobalamin II (holo-TC-II) and holo-haptocorrin (holo-TC1 and III) in serum by adsorbing holo-TC-II on microfine silica. J Nutr Biochem 1991;2:455–463.
27 Kiossoglou, KA, Mitus WJ, Dameshek W: Chromosomal aberrations in pernicious anemia. Study of three cases before and after therapy. Blood 1965;5:662–682.
28 Heath CW: Cytogenetic observations in vitamin B_{12} and folate deficiency. Blood 1966;27: 800–815.
29 Das KC: Nutritional megaloblastosis. Symp Nutr Anaemias. Proc Second Congr Asian Pacific Div, Int Soc Haematol, New Delhi, 1976, pp 14–15.
30 Hoffbrand AV, Pegg AE: Base composition of normal and megaloblastic bone marrow DNA. Nat New Biol 1972;235:187–188.
31 Khobta A, Carlo-Stella C, Capranico G: Specific histone patterns and acetylase/deacetylase activity at the breakpoint-cluster region of the human MLL gene. Cancer Res 2004;64:2656–2662.
32 Zhao J: Coordination of DNA synthesis and histone gene expression during normal cell cycle progression and after DNA damage. Cell Cycle 2004;3:695–697.
33 Norbury CJ, Zhivolovsky B: DNA damage-induced apoptosis. Oncogene 2004;23:2797–2808.
34 Wang C, Fu M, Pestell RG: Histone acetylation/deacetylation as a regulator of cell cycle gene expression. Methods Mol Biol 2004;241: 207–216.
35 McLaughlin F, La Thangue NB: Histone deacetylase inhibitors open new doors in cancer therapy. Biochem Pharmacol 2004;68: 1139–1144.
36 Ingram CF, Davidoff AN, Marais F, Sherman GG, Mendelow BV: Evaluation of DNA analysis for evidence of apoptosis in megaloblastic anaemia. Br J Haematol 1997;96:576–583.
37 Koury MJ, Price JO, Hicks GG: Apoptosis in megaloblastic anemia occurs during DNA synthesis p53-independent, nucleoside-reversible mechanism. Blood 2000;96:3249–3255.
38 Gu L, Wu J, Qiu L, Jennings CD, Li GM: Involvement of DNA mismatch repair infolate deficiency-induced apoptosis small star, filled. J Nutr Biochem 2002;13:355–363.
39 Li GM, Presnell SR, Gu I: Folate deficiency, mismatch repair-independent apoptosis and human disease. J Nutr Biochem 2003;14:568–575.
40 Das KC, Easow SK, Jadaon M, Das M, Mohanty D: The development of megaloblastosis is a species-restricted phenomenon. 9th Annu Health Sci Poster Conf, Kuwait, April 19–21, 2004, p 284.

Review

Medical Principles and Practice

Med Princ Pract 2005;14(suppl 1):15–19
DOI: 10.1159/000086180

Received: April 20, 2004
Revised: April 19, 2005

Sickle Cell Disease: A Multigenic Perspective of a Single-Gene Disorder

Abdullah Kutlar

Medical College of Georgia, Augusta, Ga., USA

Key Words
Sickle cell disease · Genetic polymorphisms

Abstract
Although significant progress has been made in our understanding of sickle cell disease (SCD) and in the development of new therapies, many questions are still unanswered, and a cure remains elusive. This is particularly evident in the clinical heterogeneity of the disease. Studies have shown the importance of high hemoglobin F determinants and α-thalassemia as modifiers of disease severity, but these alone do not explain the diversity that is seen. This paper focuses on recent advances on the effect of nonglobin genetic modifiers on the SCD phenotype. The roles of polymorphic variants of (1) methylenetetrahydrofolate reductase gene in the pathogenesis of avascular necrosis, (2) factor V R485K and risk of venous thrombosis, and (3) UDP glucuronosyltransferase-1 polymorphism on serum bilirubin levels in SCD are discussed. Mention is made of genetic polymorphisms that might predispose to stroke. The application of gene expression profiling to the study of SCD is very promising and some preliminary data are provided.

Introduction

Almost half a century has elapsed since the elucidation of the molecular basis of the sickle hemoglobin mutation in the mid 1950s. Although significant progress has been made in our understanding of the disease and in the development of new therapies, many questions are still unanswered, and a cure remains elusive. This is particularly evident by the clinical heterogeneity of sickle cell disease (SCD), which ranges from a mild clinical course with survival into the 6th and 7th decades to a very severe presentation with significant organ damage and death at relatively young ages [1].

Most of the advances in unraveling the phenotypic heterogeneity of SCD have been made in the past 30 years, with the dawn of the molecular era. Studies in this period have clearly shown the importance of high hemoglobin F (Hb F) determinants and α-thalassemia as modifiers of disease severity. The impact of these modifiers has been well established and will not be discussed in this presentation [1]. We will rather try to summarize recent advances on the effect of nonglobin genetic modifiers on the phenotype of SCD. This process has been greatly facilitated by the completion of the Human Genome Project, which has identified a large number of single nucleotide polymorphisms (SNPs) in and around many genes that can potentially impact different aspects of SCD. The realization that α-thalassemia and variation in Hb F levels alone do not account for the vast clinical diversity of

KARGER

Fax +41 61 306 12 34
E-Mail karger@karger.ch
www.karger.com

1011–7571/05/0147–0015$22.00/0

Accessible online at:
www.karger.com/mpp

Prof. Abdulla Kutlar
Sickle Cell Center, Medical College of Georgia
1521 Pope Avenue
Augusta, GA 30912 (USA)
Tel. +1 706 721 2171, Fax +1 706 721 4575, E-Mail akutlar@mail.mcg.edu

the disease has provided further impetus to search for other genetic modifiers of SCD.

The well-known consequences of the sickle mutation and its downstream effects are (1) a chronic hemolytic anemia, (2) episodic vaso-occlusion with resultant characteristic painful episodes, and (3) chronic organ damage. The clinical phenotype of SCD is highly variable; complications such as acute chest syndrome, stroke, leg ulcers, and avascular necrosis (AVN) do not occur in all of the patients, suggesting that factors beyond the β6 Glu→Val mutation play a role in the pathogenesis of these syndromes. Recent work, particularly by Hebbel and Vercellotti [2], has established that SCD is an 'inflammatory state' and that the endothelium is activated. Thus, the emerging view of the pathogenesis of many complications of SCD involves complex interactions between sickle reticulocytes, neutrophils, monocytes, and the endothelium. It therefore follows that many factors that are important in the pathways leading to inflammation, cell adhesion, and coagulation will be important in the pathophysiology of the disease and that variations in the expression of a variety of molecules in these pathways will contribute to the heterogeneity of SCD.

The approach to the study of this complexity in the postgenome era should involve several different methodologies including: (1) analysis of polymorphisms in 'candidate genes' by high throughput methods and the association of these polymorphisms with distinct phenotypic features of the disease, (2) the study of differential gene expression in various tissues using cDNA microarrays, and (3) the application of proteomics. The tissues and cells that are particularly important in this regard include blood cells (neutrophils, monocytes, reticulocytes), bone marrow, endothelial cells, and liver tissues. A candidate gene approach should include a large number of patients with and without a certain complication (e.g. stroke) to determine the role of these polymorphisms as risk factors. Pooling studies to discover the role of genome-wide SNPs as risk factors for certain complications of the disease is also emerging as a viable approach. In all of these studies, careful characterization of the phenotype remains of critical importance. A study of sickle cell patients from different populations is likely to yield important information because of the differences of the genetic backgrounds in these populations and their potential implications on the disease phenotype. The important SCD-related 'phenotypes' should include complications such as stroke and CNS disease, frequent vs. infrequent vaso-occlusive episodes (VOEs), frequency of the acute chest syndrome, pulmonary hypertension, AVN of the hips and shoulders, gallbladder disease and cholecystectomy, leg ulcers, renal involvement, priapism, and retinopathy.

For the past several years, our center and many other centers have focused on the study of nonglobin genetic modifiers of sickle cell disease. Some of the findings of these early studies will be summarized below.

AVN and Methylenetetrahydrofolate Reductase C677T Mutation

Studies of large patient populations such as the Cooperative Study of Sickle Cell Disease have shown that co-inherited α-thalassemia, age, high hematocrit, and frequent VOEs are risk factors for the development of AVN of the femoral head in SCD [3]. We determined the frequency of the thermolabile methylenetetrahydrofolate reductase (MTHFR) C677T mutation in our sickle cell population, because of its association with elevated serum homocysteine levels and resultant vascular complications. Overall, the MTHFR mutation was present in 16% of the sickle cell patients in our center (1.8% homozygote). There was a strong association of the presence of MTHFR with AVN, with 35.6% of the AVN patients having the MTHFR mutation as opposed to only 12.9% of sickle cell patients without AVN ($p = 0.006$). Furthermore, the presence of concomitant α-thalassemia and MTHFR was found to be additive in terms of the risk of developing AVN [4]. This association, however, was not confirmed in the high Hb F population of Kuwaiti sickle cell patients, suggesting that different genetic factors may be operative in different populations [5].

Factor V R485K Polymorphism

First described as a neutral polymorphism in the factor V gene in nucleotide 1628 in exon 10 by Gandrille et al. [6], the Arg→Val substitution at residue 485 was found at high frequency (32.4%) in sub-Saharan Africans. This polymorphism was later associated with a risk of thrombosis and coronary artery disease in Asian populations and reported to lead to a mild activated protein C resistance [7]. A study of the frequency of this polymorphism in our sickle cell population showed a similar allele frequency of 0.29 (45.1% heterozygous, 6.6% homozygous). Although no association of this polymorphism has been established with certain complications of SCD (frequent VOEs, acute chest syndrome, leg ulcers, AVN, and priapism), interestingly, the presence of this mutation has been shown to be associated with an increased risk of central venous catheter-associated thrombosis (p = 0.006, odds ratio = 4.9; table 1) [8].

UGT1A1 Polymorphism and Bilirubin Levels

The gene UGT1A1 encodes the enzyme UDP glucuronosyltransferase-1 that mediates the glucuronidation of bilirubin. A common polymorphism in the TATA box of this gene was found to be associated with a decreased enzyme activity and indirect hyperbilirubinemia (Gilbert's syndrome). The sequence A(TA)6TAA is presumed to be the wild type while A(TA)7TAA is associated with Gilbert's syndrome. The coinheritance of this (TA)7 polymorphism with hereditary hemolytic anemias, such as hereditary spherocytosis, α-thalassemia, and G6PD deficiency, results in marked hyperbilirubinemia and increased incidence of gallstones [9]. We, and others, have studied the frequency of this polymorphism in the sickle cell population and its impact on bilirubin levels and incidence of gallstones and cholecystectomy [10, 11]. In all of these studies, an association was found between the (TA)7 genotype and bilirubin levels; those with 7/7 had the highest mean bilirubin levels compared to the genotypes with 6/6 and 6/7. Our results are summarized in table 2. Furthermore, the impact of the 7/7 genotype on treatment with hydroxyurea has recently been reported by Henney et al. [12]. These authors reported a normalization of the bilirubin levels in sickle cell patients with the 6/6 genotype upon treatment with hydroxyurea in contrast to those with 7/7 and 6/7 genotypes whose bilirubins failed to fall to normal levels.

Table 1. R485K polymorphism and catheter-induced thrombosis in sickle cell patients

R485K	Thrombosis (n = 10)	Control (n = 10)
+/+	6	2
–/+	3	2
–/–	1	6

p = 0.006; odds ratio = 4.9.

Table 2. UGT1A1 polymorphism and bilirubin levels in steady-state SCD

	7/7 (n = 22)	6/7 (n = 35)	6/6 (n = 10)
Reties, %	10.9	11.4	11.3
Hb F, %	6.6	11.0	8.4
Total bilirubin level, mg/dl	5.6	3.8	3.8
Cholecystectomy, %	59	51	60

Genetic Polymorphisms as Risk Factors for Stroke

Clinical stroke occurs in ~11% of US sickle cell patients by 20 years of age [13]. Increased velocities of blood flow (>200 cm/s) as measured by transcranial Doppler (TCD) ultrasound in major intracranial arteries has been shown to be a predictor of stroke risk in children with SCD [14]. The value of TCD as a surrogate marker of cerebrovascular disease and a predictor of stroke risk has been validated in the randomized STOP trial, and a therapeutic intervention based upon this risk stratification (prophylactic transfusions in children with high TCD velocities) has been proven to reduce the stroke risk [14]. Despite the validation of the utility of TCD as a clinical tool for predicting stroke risk, biologic factors underlying the development of cerebrovascular disease and stroke in the SCD population are poorly understood. The presence of α-thalassemia has been shown to be protective of cerebrovascular disease and stroke [13]. Several recent studies have sought an association between certain genetic polymorphisms and stroke risk in SCD. Driscoll et al. [15] found a higher rate of stroke in siblings with SCD (p = 0.0012), suggesting the role of genetic factors. A recent study has found an association with HLA genotypes and stroke; of particular interest is the association of certain genotypes with distinct subtypes (large vessel vs. small vessel stroke) in a retrospective study of the Cooperative Study of Sickle Cell Disease population [16]. Another recent study [17] reported a protective role for VCAM G1238C polymorphism against symptomatic stroke (odds ratio = 0.35, p = 0.04). An ongoing study at our center is looking at the association of high TCD velocity with 28 genetic polymorphisms among 25 candidate genes in 630 patients (230 high TCD, 400 normal TCD) enrolled in the STOP and STOP II trials. The candidate genes include those associated with coagulation and thrombophilia (factor V, factor VII, factor XIII, prothrombin, thrombomodulin, fibrinogen, PAI-1, MTHFR), endothelial cell function and inflammation

(VCAM-1, ICAM-1, selectins, TNF-α, Apo A and Apo E), platelet function and activation (GpIIb/IIIa, Gplb IX-V, Gpla/IIIa), and vascular reactivity (ACE). The analysis of polymorphisms will be accomplished by a high throughput SNP genotyping method using the MALDI-TOF-based Massarray™ system (Sequenom, San Diego, Calif., USA).

Gene Expression Profiling

Gene expression profiling utilizing cDNA microarrays is a relatively novel method, particularly in terms of its applications to the study of SCD. Few published reports exist in this area. These include the study of gene expression profiling from the kidney tissue in transgenic sickle cell mice [18], profiling of bronchial endothelial cells in patients with SCD and acute chest syndrome [19], and a study of differential gene expression in peripheral blood cells from SCD patients with differing vaso-occlusive crisis frequency [20].

The quality of the data obtained from microarray studies depends upon several factors: (1) a meticulous study design, particularly a clear and unambiguous definition of the 'phenotype(s)' to be studied, (2) elimination of technical/methodological pitfalls (e.g. quantity/quality of the RNA, methods of specimen collection and processing, conditions and timing of hybridization, and elimination of intrapatient variation by repeated sampling), and (3) expertise in bioinformatics in the interpretation of vast amounts of data generated.

In patients with SCD, the tissues that may be amenable to gene expression profiling studies include peripheral blood cells, endothelial cells (circulating endothelial cells, as well as endothelial cells from various tissues), bone marrow, and liver tissue. We have conducted preliminary studies of gene expression with commercially available cDNA microarrays (Affymetrix U95, Affymetrix, Santa Clara, Calif., USA) from peripheral blood neutrophils and reticulocytes in SCD patients with discordant rates of painful episodes in an effort to elucidate the factors contributing to different crisis frequency. Two patients who were quite similar in terms of demographic (age, gender, ethnic/racial background) and hematologic (hemoglobin, reticulocyte count, neutrophil count, Hb F level, α-globin genes and β^S haplotypes) characteristics, but very different in terms of annual crisis frequency (0.75/year vs. 6/year) were chosen for the preliminary study. An analysis of gene expression patterns from purified peripheral blood neutrophils revealed that of the 7,070 genes represented on the microarray, 536 had a >2-fold expression difference between the 2 patients. Sixteen genes had a higher expression and 27 genes had a lower expression in the patient with a high crisis frequency as compared to the patient with a lower vaso-occlusive crisis frequency. The genes with a higher expression in the patient with frequent VOEs included CD11b/CD18, phospholipase A_2, C and D, and IL-8, suggesting a higher level of neutrophil activation in this patient.

The application of the microarray technology to the study of SCD is still in its infancy. Although the preliminary data look interesting, these results need to be confirmed and validated through a large number of experiments in a large number of patients. The available data appear promising and show the feasibility of the application of this technology to SCD.

Conclusions

The development of high throughput genotyping methods, the microarray technology, and the completion of the Human Genome Project have already begun to revolutionize medicine and biology. The application of these techniques to the study of SCD is in its very early stages and is likely to yield novel information.

A candidate gene approach and study of genome-wide SNPs and their association with certain complications of SCD is expected to further clarify the basis of clinical/phenotypic heterogeneity of this single-gene disorder. A study of gene expression profiling coupled with proteomics will shed further light on the functional aspects and differences between patients. This information will not only lead to a better understanding of the pathogenesis and pathophysiology of many complications of the disease, but will also likely result in the identification of novel therapeutic targets and discovery of new genes with prognostic and therapeutic implications.

References

1 Steinberg MH, Rodgers GP: Pathophysiology of sickle cell disease: Role of cellular and genetic modifiers. Semin Hematol 2001;38:299–306.
2 Hebbel RP, Vercellotti GM: The endothelial biology of sickle cell disease. J Lab Clin Med 1997;129:288–293.
3 Milner PF, Kraus AP, Sebes JI, Sleeper LA, Dukes KA, Embury SH, Bellevue R, Koshy M, Moohr JW, Smith J: Sickle cell disease as a cause of osteonecrosis of the Femoral Head. N Engl J Med 1991;325:1476–1481.
4 Kutlar A, Kutlar F, Turker I, Tural C: The methylene tetrahydrofolate reductase (C677T) mutation as a potential risk factor for avascular necrosis in sickle cell disease. Hemoglobin 2001;25:213–217.
5 Adekile AD, Kutlar F, Haider MZ, Kutlar A: Frequency of the 677C→T mutation of the methylenetetrahydrofolate reductase gene among Kuwaiti sickle cell disease patients. Am J Hematol 2001;66:263–266.
6 Gandrille S, Greengard JS, Alhenc-Gelas M, Juhan-Vague I, Abgrall JF, Jude B, Griffin JH, Aiach M: Incidence of activated protein C resistance caused by the ARG 5065 GLN mutation in factor V in 113 unrelated symptomatic protein C-deficient patients. Blood 1995;86: 219–224.
7 Le W, Yu JD, Tao R, You B, Cai X, Cao WJ, Huang W, He RM, Zhu DL, Chen Z, Gong LS: Association of the R485K polymorphism of the factor V gene with poor response to activated protein C and increased risk of coronary artery disease in the Chinese population. Clin Genet 2000;57:296–303.
8 Ustun C, Adams GT, Kutlar F, Elam D, Clair B, Daitch L: Factor V R485K polymorphism is a risk factor for catheter-induced thrombosis in sickle cell disease. 30th Anniversary of the National Sickle Cell Disease Program and the SCDAA, Washington, September 17–21, 2002.
9 Sampietro M, Lipica L, Perrero L, Comino A, Martinez di Montemuros F, Cappellini MD, Fiorelli G: The Expression of uridine diphosphate glucuronosyltransferase gene is a major determinant of bilirubin level in heterozygous β-thalassaemia and in glucose-6-phosphate dehydrogenase deficiency. Br J Haematol 1997; 99:437–439.
10 McKie K, Kutlar F, Sromek E, Litaker M, Woods KF, Kutlar A: Uridine diphosphate glucuronosyl transferase-1 (UGTIAI) promoter polymorphism and bilirubin levels in patients with sickle cell disease. Blood 1999; 94(suppl 1):861.
11 Passon RG, Howard TA, Zimmerman SA, Schultz WH, Ware RE: Influence of bilirubin uridine diphosphate-glucuronosyltransferase IA promoter polymorphisms on serum bilirubin levels and cholelithiasis in children with sickle cell anemia. J Pediatr Hematol Oncol 2001;23:448–451.
12 Henney MW, Howard TA, Zimmerman SA, Ware RE: UGT1A promoter polymorphism influence bilirubin response to hydroxyurea therapy in sickle cell anemia. J Lab Clin Med 2003;141:279–282.
13 Ohene-Frempong K, Weiner SJ, Sleeper LA, Miller ST, Embury S, Moohr JW, Wethers DL, Pegelow CH, Gill FM: Cerebrovascular accidents in sickle cell disease: rates and risk factors. Blood 1998;91:288–294.
14 Adams RJ, McKie VC, Hsu L, Files B, Vichinsky E, Pegelow C, Abboud M, Gallagher D, Kutlar A, Nichols FT, Bonds DR, Brambilla D: Prevention of a first stroke by transfusion in children with sickle cell anemia and abnormal results on transcranial Doppler ultrasonography. N Engl J Med 1998;339:5–11.
15 Driscoll MC, Hurlet A, Styles L, McKie V, Files B, Olivieri N, Pegelow C, Berman B, Drachtman R, Patel K, Brambilla D: Stroke risk in siblings with sickle cell anemia. Blood 2003;101:2401–2404.
16 Hoppe C, Klitz W, Noble J, Vigil L, Vichinsky E, Styles L: Distinct HLA associations by stroke subtype in children with sickle cell anemia. Blood 2003;101:2865–2869.
17 Taylor VI JG, Tang DG, Savage SA, Leitman SF, Heller SI, Serjeant GR, Rodgers GP, Chanock SJ: Variants in the VCAM1 gene and risk for symptomatic stroke in sickle cell disease. Blood 2002;1000:4303–4309.
18 Nagel RL, Fabry ME, Kaul DK, Rybicki AC: Up-regulation of gene expression in the chronically damaged kidney in sickle transgenic mice. Blood 2002;100(suppl):2607.
19 Adewoye A, Safaya S, Frampton G, Lenburg M, Klings EX, Odhiambo A: Pulmonary artery endothelial cells exposed to acute chest syndrome plasma express a novel repertoire of genes. Blood 2002;100(suppl):1756.
20 Kutlar A, Kutlar F, Clair B, Nechtman J, Joshi RM, Daitch L: Gene expression profiling of peripheral blood cells from sickle cell patients with differing crisis rates. Blood 2002;100(suppl):3549.

Review

Medical Principles and Practice

Med Princ Pract 2005;14(suppl 1):20–26
DOI: 10.1159/000086181

Received: June 6, 2004
Revised: November 9, 2004

Molecular Techniques in the Diagnosis and Management of Infectious Diseases: Do They Have a Role in Bacteriology?

Ian R. Poxton

Division of Medical Microbiology, Centre for Infectious Diseases, University of Edinburgh College of Medicine and Veterinary Medicine, Edinburgh, UK

Key Words

Diagnostic methods, traditional versus molecular techniques · Resistance to antimicrobial agents · Koch's postulates

Abstract

The diagnosis and management of bacterial diseases has been done by traditional methods for a century or more. With the advent of molecular methods, however, these traditional approaches are being challenged. This review examines the pros and cons of traditional versus modern methods and tries to answer the question: when are molecular methods useful or essential? The following topics are addressed with appropriate examples: diagnosis; identification, typing and fingerprinting; pathogenesis; patient management; susceptibility to disease, and resistance to antimicrobial agents. It was concluded that there is still a place for both traditional and modern molecular methods, and training of staff must include both methodologies. Innovation is encouraged – but new technologies must be thoroughly tested before introduction into the routine lab. Liaison between laboratory scientist and physician is important, but above all experience is paramount.

Introduction

In microbiology, a commonly held view is that virologists have to use molecular techniques to diagnose and manage disease, yet bacteriologists do not need to use them as bacteria can be easily cultured and identified. After all, traditional means have been used for over a hundred years. However, is this view really true in the 21st century?

The advantages of traditional culture methods are obvious in that they show the 'whole picture', they are quantitative, isolates can be typed or fingerprinted (by phenotypic or genotypic methods), antibiotic sensitivity tests can be done, isolates can be archived and the methods have stood the test of time. The disadvantages are that they are slow, require skilled, experienced personnel and only cultivable bacteria can be detected, which is not possible if the culture medium does not support their growth or if antibiotics have been used. Furthermore, it is questionable whether or not they are actually quantitative.

Similarly, there are both advantages and disadvantages of molecular methods. Advantages include their speed, increased sensitivity and the fact that they can be done on inadequately stored specimens. In addition, databases and phylogenetic trees can be readily made and automation is possible. Disadvantages are that significant costs

KARGER

Fax +41 61 306 12 34
E-Mail karger@karger.ch
www.karger.com

1011–7571/05/0147–0020$22.00/0

Accessible online at:
www.karger.com/mpp

Prof. Ian R. Poxton
Division of Medical Microbiology, Centre for Infectious Diseases
University of Edinburgh College of Medicine and Veterinary Medicine
Edinburgh (UK)
Tel./Fax +44 131 650 3128, E-Mail i.r.poxton@ed.ac.uk

are involved (in reagents, apparatus, personnel), no bacteria are isolated for further studies, specimen adequacy cannot be determined, there is a risk of contamination especially if done by untrained staff (non-microbiologists), confirmatory tests are often necessary and the methods are not yet widely accepted for medicolegal testing.

Diagnostic Methods

With regard to diagnosis, one of the first considerations is whether the method must be rapid or not. Although speed is often important, it is sometimes appropriate and in the interest of both the patient and the economics of the process for a result to be ready in hours or days rather than minutes or hours. Slower and often cheaper procedures are more appropriate when an immediate result is not deemed necessary. A rapid result is required if therapy or management can be instigated only after the result is known. However, empirical treatment with a broad-spectrum antibiotic may be used to buy time.

Without doubt, the sooner a result is known the better for life-threatening infections such as meningitis or septicaemia. Blood culture methodology has advanced greatly, but results still take time. Similarly many paediatric infections require an early diagnosis. From a different point of view, diagnosis of sexually transmitted infections can benefit from a rapid diagnosis, so that patients can be treated before they leave the clinic.

The early generation of rapid tests was based on detection of antigens in specimens, utilising antigen capture in enzyme immunoassay, or antigen detection by immunofluorescence. The more recent tests involve nucleic acid methods, relying on detection of specific genes by an amplification method such as PCR, including the much more quantitative real-time PCR, or the use of gene probes. A special case where molecular methods are making huge inroads is for the rapid diagnosis of tuberculosis and other mycobacterial infections. *Mycobacterium tuberculosis* and related species are extremely slow growers in laboratory media and take several weeks to culture. Proper diagnosis is needed both to institute appropriate therapy and also to isolate patients to prevent transmission of disease.

A comparison of traditional culture with modern, rapid tests is illustrated in table 1. Here the diagnosis of *Chlamydia trachomatis* infection is examined [1]. The specificity of all methods is fairly high, but the sensitivity for the immunoassays is sometimes poor.

Table 1. Sensitivity and specificity of *C. trachomatis* testing

Method	Sensitivity, %	Specificity, %
Culture	70–90	100
Nucleic acid amplification	90–100	97–100
DNA probe	67–96	96–100
Immunofluorescence	70–100	>95
Enzyme immunoassay	43–92	92–100

Modified from Medscape Molecular Medicine [1].

For routine diagnostic procedures, a combination of both approaches, traditional and molecular, can probably be employed for optimal information. To illustrate this combined approach, the following is an example of a typical investigation of a stool specimen for a suspected infection caused by an enterohaemorrhagic *Escherichia coli,* an easily cultured bacterium. Following clinical symptoms of bloody diarrhoea and haemolytic uraemic syndrome/thrombocytopaenia purpura, the stool specimen is cultured on MacConkey's agar containing 1% D-sorbitol. Non-sorbitol-fermenting colonies are confirmed as *E. coli* O157 by latex agglutination tests. In an accompanying epidemiological investigation, enrichment might be done from food, water or faeces by immunomagnetic beads. The *stx* gene, and possibly LEE or other virulence genes could be detected with DNA probes. Finally pulsed-field gel electrophoresis might be used to compare isolates.

Identification, Typing and Fingerprinting – Epidemiological or Taxonomic Tools

The use of molecular methods for bacterial identification is most useful for bacteria that are either difficult to grow or are easy to grow yet are biochemically unreactive.

Examples of the latter are members of the genus *Actinomyces.* Val Hall et al. [2, 3] have made major improvements to the identification of these species using a technique based on amplified 16S ribosomal DNA restriction analysis.

Investigation of slow or non-cultivable bacteria has usually been done using traditional serology. Now such organisms can be investigated by probe or PCR detection, and new associations between symptoms and organism can be drawn. An important example of such an investi-

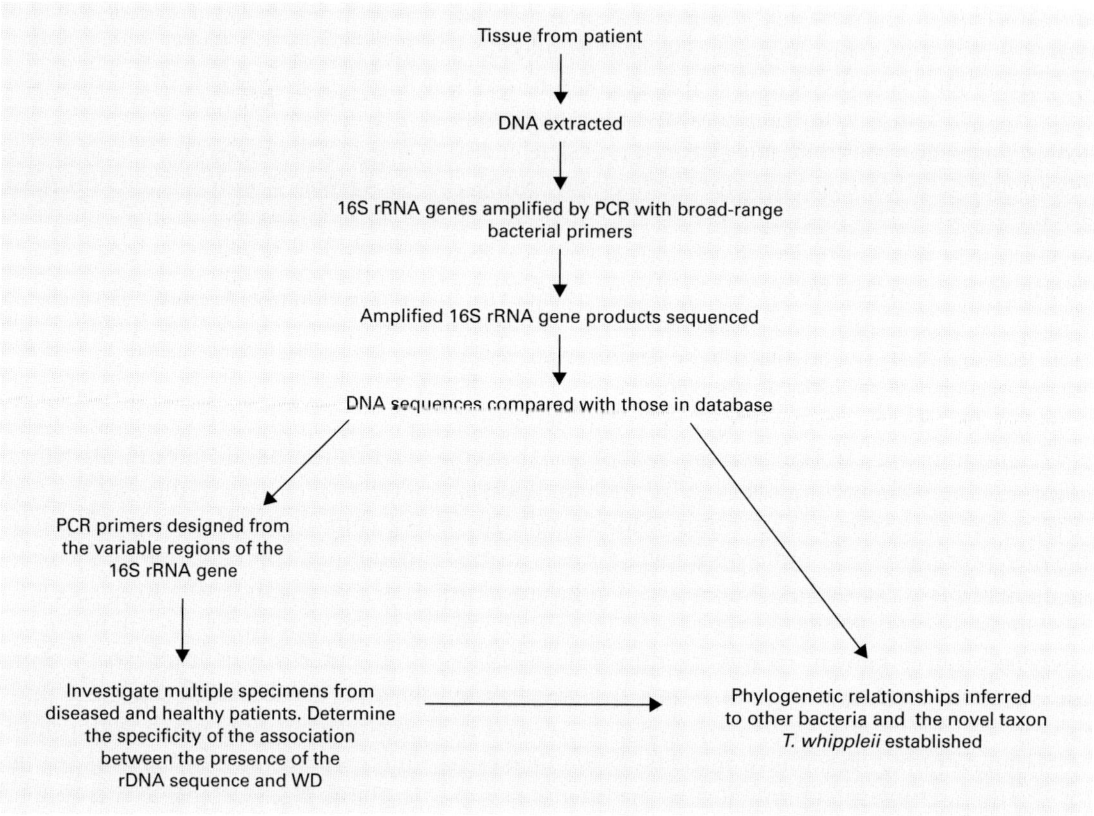

Fig. 1. The investigation of Whipple's disease (WD; modified from Fredricks and Relman [6]).

gation is that of Whipple's disease. This disease is a rare infection of the small intestine causing malabsorption. It was thought to be a bacterial disease since its discovery in 1907, since bacilli were seen in affected areas and it responded later to antibiotic treatment. The final proof that it was caused by the bacterium *Tropheryma whippelii* was done entirely by a sequence-based approach [4, 5]. The approach is summarised diagrammatically in figure 1, modified from Fredricks and Relman [6].

Studying Pathogenesis

The aetiology of an infectious disease has traditionally been investigated through the establishment of Koch's postulates. However, for many reasons, not least those of ethics, inability to culture an organism and the preponderance of opportunist infections arising from the endogenous microbiota, they have been impossible to fulfil. As a result of this, molecular Koch's postulates have been created [7], which are listed in table 2.

Table 2. Establishment that a bacterium is the cause of a disease and identification of virulence factors by fulfilment of 'molecular Koch's postulates'

1	Identify a gene coding for a virulence determinant that is associated with all bacteria that cause the disease and is not present in avirulent strains
2	Inactivation of the gene coding for the virulence factor should substantially decrease pathogenicity
3	Replacement of the mutated gene with the normal wild-type gene should restore pathogenicity fully
4	The gene should be expressed during the infection and disease process
5	A protective immune response should be produced to the gene product

Patient Management

Monitoring the progression of disease is done traditionally by clinical monitoring, culture and measuring a serological response. However, with the advent of molecular methods, such traditional approaches may be superseded. For example, detection of mediators of disease is traditionally accomplished by clinical measurements, from simple temperature changes to acute physiology scores. Measurement of acute-phase proteins such as C-reactive protein has long been used [8]. Other mediators are proinflammatory cytokines such as tumour necrosis factor alpha (TNF-α), interleukin (IL)-1, IL-2, IL-6, IL-8 and IFN, which are usually detected by immunoassays or bioassays. However, molecular approaches are now possible, for example by detecting specific messenger RNAs by reverse transcriptase PCR [9, 10].

Another method, which is 'molecular' but not nucleic acid-based, involves the detection of the prime mediator of inflammation: endotoxin [bacterial lipopolysaccharide (LPS)]. Traditionally, endotoxin has been detected by measuring its pyrogenicity in the rabbit fever model. More recently, this has been largely superseded by the *Limulus* amoebocyte lysate assay, which is an assay based on the exquisite sensitivity of the lysed amoebocytes from the horseshoe crab *L. polyphemus.* Exposure to endotoxin results in gelation of the lysate. Either the original clotting (gelation) assay or the more specific colourimetric version can be used. Despite years of experience, it is not routinely used in diagnosis, mainly because of the difficulty in use, especially its lack of reproducibility probably due to the relatively short half-life of endotoxin in plasma. Downstream mediators such as IL-6 have a much greater half-life. Also the extreme sensitivity of the test makes it very prone to contamination, which can cause false positives.

An indirect method for endotoxin determinations has been used a great deal in our laboratory and elsewhere. In this assay, EndoCAb™, antibodies to endotoxin core are measured and their levels are inversely related to endotoxin level over time. It is also a prognostic indicator, where high levels are considered protective for endotoxin-related problems, as discussed below. A kit is available commercially (Hycult Biotechnology BV, The Netherlands). The coating (detecting) antigens in this ELISA are incomplete for core-lipid A structures from R-mutants of LPSs: from *E. coli, Salmonella minnesota, Klebsiella pneumoniae* and *Pseudomonas aeruginosa* combined in equimolar proportions and complexed with polymyxin. This assay detects antibodies that cross-react with the common epitopes expressed by these incomplete core LPSs in blood donors and patients. Early investigations employing this assay were reviewed by Barclay in 1995 [11], and since then the kit has been used for many different investigations [12–14].

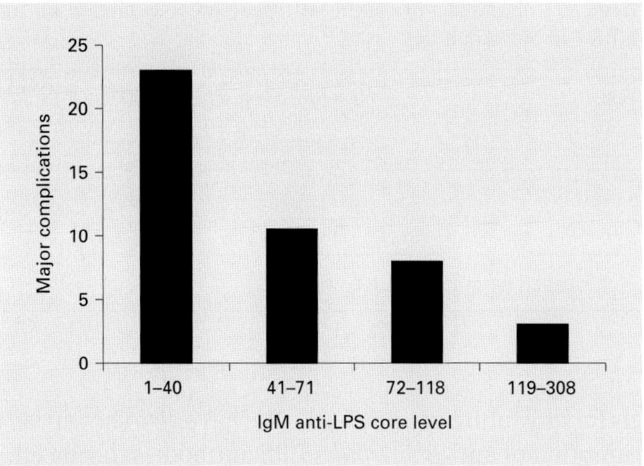

Fig. 2. High pre-operative IgM levels protect against endotoxaemia following cardiac surgery. This figure shows that pre-operative low serum anti-endotoxin core antibody (EndoCAb) levels are a predictor of adverse outcome following cardiac surgery. Major complication is defined as either in-hospital death or postoperative length of stay greater than 10 days. Antibody levels are in median units for the EndoCA ELISA (modified from Bennett-Guerrero et al. [17]).

Investigating Susceptibility to Infectious Diseases

Disease susceptibility is often associated with the genotype of the individual. Polymorphisms in several different genes have been associated with differential susceptibility to infections and the magnitude of symptoms. Polymorphisms in cytokine genes such as TNF-α and Toll-like receptors are well-cited examples. Recently, however, doubt has been cast on how important these polymorphisms are. Bayley et al. [15] have questioned the relevance of TNF polymorphisms and suggested that the evidence indicates that they are important only in malaria. Similarly, we have questioned the significance of Toll-like receptor 4 polymorphisms [16].

As mentioned above, deficiencies in antibodies to endotoxin, as measured by the EndoCAb method, indicate possible susceptibility to endotoxin-related problems. Conversely, high levels of antibodies are considered to be protective. Studies show that if patients have high pre-operative levels of antibodies to the LPS core, they have a much better chance of not developing endotoxin-related complications than patients who have low levels [17, 18]. Figure 2 summarises the results of the study conducted on cardiac surgery patients [17] demonstrating that high pre-operative levels of antibodies to the LPS core afford protection, and these high levels should be the target lev-

Table 3. Comparison of culture with gene probe detection kit for MRSA in blood cultures

Test	Comparison with *mecA/femB* PCR	
	sensitivity, %	specificity, %
EVIGENE	100	99.5
Culture	88.2	99.5

Modified from Levi and Towner [22].

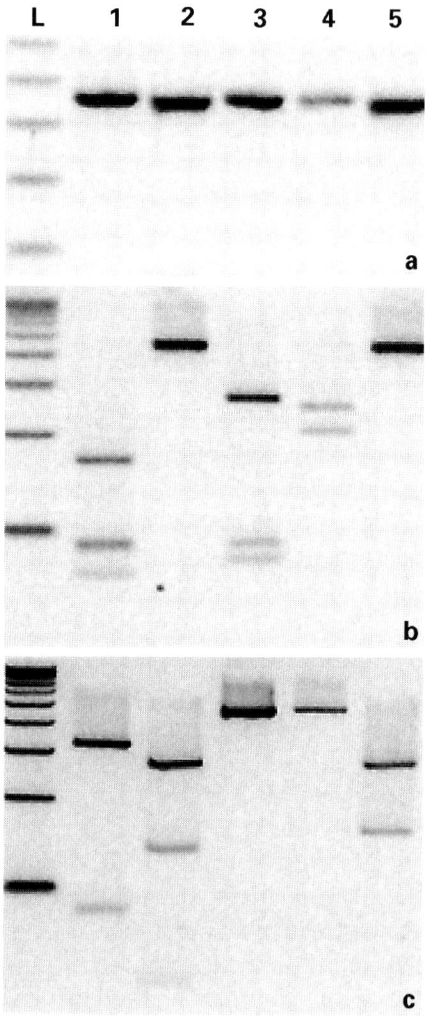

Fig. 3. PCR-RFLP analyses for identification and characterization of *nim* resistance genes of *Bacteroides* spp. PCR products (**a**) and RFLP profiles for *Hpa*II (**b**) and *Taq*I (**c**) for genes *nimA* to *nimE*. L = 100-bp ladder. From Stubbs et al. [24], with permission from A.S.M. *(Journal of Clinical Microbiology).*

els for any future vaccine trials. We have also shown consumption of anti-*Bacteroides* LPS antibodies during episodes of sepsis, suggesting that the Gram-negative anaerobes, as well as other gastro-intestinal Gram-negative bacteria may be translocated into the peripheral circulation during sepsis [19].

We have recently developed and tested preclinically a vaccine designed to boost natural levels of anti-LPS core antibodies [20, 21]. The vaccine contains a set of R-LPSs in liposomes. It is non-pyrogenic, gives good seroconversion and is cross-reactive with smooth LPSs in mice and rabbits, and is protective in mouse models. New work is concentrating on core oligosaccharide-protein conjugates as vaccines and results are similar to the liposomal vaccines.

Detecting Resistance to Antimicrobial Agents

Several molecular methods have been developed to detect resistance to antimicrobial agents by detecting resistance genes. These methods certainly have a role in investigating the epidemiology of resistance genes, but when we have disk testing for cultures on primary isolation plates, MIC testing on pure cultures by broth or agar dilutions, or E-tests, are they really necessary?

Generally speaking, the answer is no, although there are a few notable exceptions which will be described here.

(a) The first example is for rapid detection of methicillin-resistant *Staphylococcus aureus* (MRSA) in blood cultures. This technique is based on probes for resistance genes and an example of this has recently been reported by Levi and Towner [22], who employed the EVIGENE™ MRSA detection kit (Statens Serum Institute, Copenhagen, Denmark). The method involves mixing samples with sterile water to lyse red blood cells. This is then incubated in broth at 37°C for 3 h. After centrifugation, the pellet is tested with the EVIGENE MRSA detection kit. The pellet is lysed by treatment with enzyme and heat, the probes *mecA, nuc,* and 16S rRNA are added in microwell strips and the results read colourimetrically in less than 3.5 h. They compared the efficacy of EVIGENE and conventional culture with *mecA/femB* PCR for detection of MRSA in 200 blood cultures, and the results compared well with culture (table 3). The time to results after positive blood culture was 7 h for EVIGENE, compared to 1+ day for conventional culture.

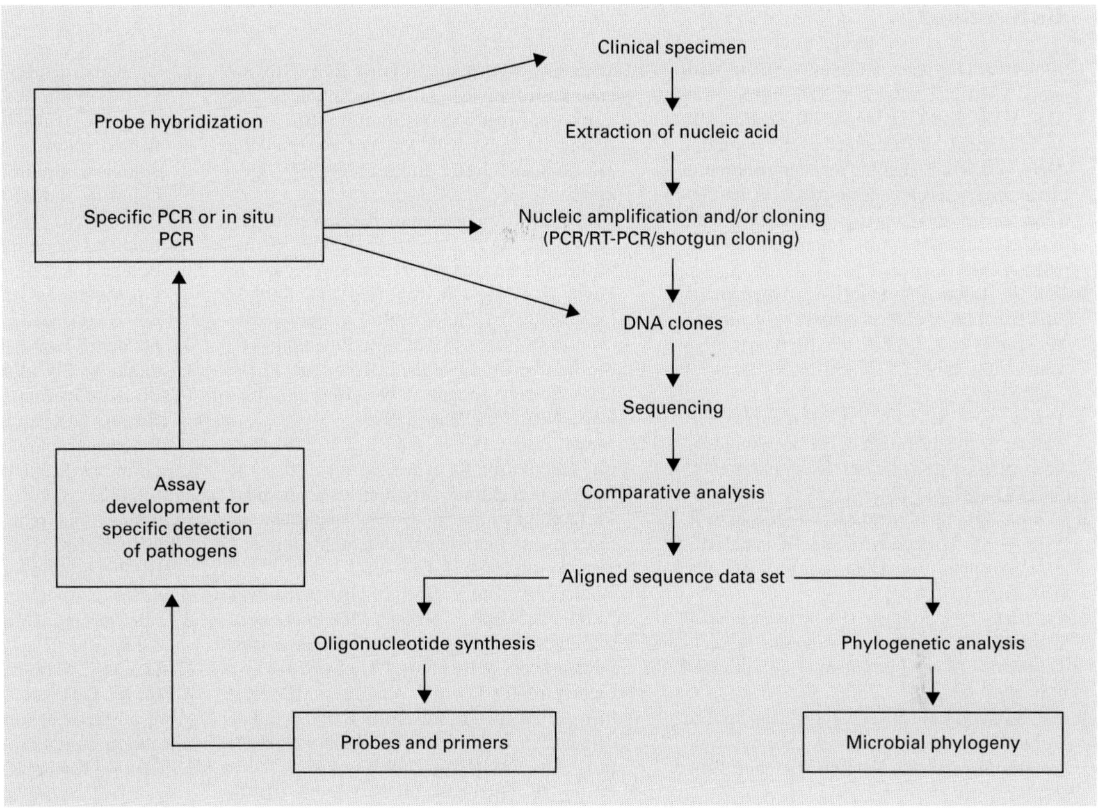

Fig. 4. Is this the way for the future? Simplified from Fredricks and Relman [6].

(b) Another example is detecting resistance to 5-nitroimidazole antibiotics (e.g. metronidazole) in anaerobic bacteria [23]. Detection of antibiotic resistance in anaerobes requires a good anaerobic technique, which is sometimes difficult to achieve. Molecular methods can be used to detect genes that code for resistance factors. In the *Bacteroides fragilis* group of species 5 *nim* genes (A, B, C, D and E) have been identified that confer resistance to 5-nitroimidazole antibiotics. Figure 3 shows the PCR-RFLP analyses of *nim* resistance genes of *Bacteroides* spp.

In summary, the detection of resistance to antimicrobial agents by molecular approaches permits very rapid detection and indicates specific resistance mechanisms, but these methods are not quantitative. MICs need to be determined separately by traditional approaches.

Figure 4 is an attempt to illustrate what the future may bring. Some laboratories are already pursuing such a scheme, although at present they are doing this in parallel with traditional methods. Innovation is to be encouraged in new technologies, but these must be thoroughly tested before introduction into the routine laboratory. Good communication is absolutely necessary between laboratory scientist and physician, and finally experience is paramount.

Conclusions

Molecular methods are advancing rapidly and are gaining an important place in all aspects of diagnosis and management of bacterial infectious diseases. However, traditional methods are still needed, and there is still much to be learned about them. The training of staff should include both methodologies.

References

1 Molecular Diagnosis of Common STDs. Medscape Molecular Medicine, 2003, vol 5.

2 Hall V, O'Neill GL, Magee JT, Duerden BI: Development of amplified 16S ribosomal DNA restriction analysis for identification of *Actinomyces* species and comparison with pyrolysis-mass spectrometry and conventional biochemical tests. J Clin Microbiol 1999;37: 2255–2261.

3 Hall V, Talbot PR, Stubbs SL, Duerden BI: Identification of clinical isolates of *Actinomyces* species by amplified 16S ribosomal DNA restriction analysis. J Clin Microbiol 2001;39: 3555–3562.

4 Relman DA, Schmidt TM, MacDermott RP, Falkow S: Identification of the uncultured bacillus of Whipple's disease. N Engl J Med 1992; 327:293–301.

5 Wilson KH, Blitchington R, Frothingham R, Wilson JA: Phylogeny of the Whipple's-disease-associated bacterium. Lancet 1991;338: 474–475.

6 Fredricks DN, Relman DA: Sequence-based identification of microbial pathogens: A reconsideration of Koch's postulates. Clin Microbiol Rev 1996;9:18–23.

7 Falkow S: Molecular Koch's postulates applied to bacterial pathogenicity – A personal recollection 15 years later. Nat Rev Microbiol 2004; 2:67–72.

8 Almirall J, Bolibar I, Toran P, Pera G, Boquet X, Balanzo X, Sauca G: Community-Acquired Pneumonia Maresme Study Group. Contribution of C-reactive protein to the diagnosis and assessment of severity of community-acquired pneumonia. Chest 2004;125:1335–1342.

9 Gerard HC, Wang Z, Whittum-Hudson JA, El-Gabalawy H, Goldbach-Mansky R, Bardin T, Schumacher HR, Hudson AP: Cytokine and chemokine mRNA produced in synovial tissue chronically infected with *Chlamydia trachomatis* and *C. pneumoniae.* J Rheumatol 2002; 29:1827–1835.

10 Ulett GC, Ketheesan N, Hirst RG: Cytokine gene expression in innately susceptible BALB/c mice and relatively resistant C57BL/6 mice during infection with virulent *Burkholderia pseudomallei.* Infect Immun 2000;68:2034–2042.

11 Barclay GR: Endogenous endotoxin-core antibody (EndoCAb) as a marker of endotoxin exposure and a prognostic indicator: A review; in Levin J, Alving CR, Munford RD, Redl H (eds): Bacterial Endotoxins: Lipopolysaccharides from Genes to Therapy. Proceedings of the 3rd Conference of the International Endotoxin Society (August 1994, Helsinki, Finland). New York, Wiley, 1995.

12 Ide M, Jagdev D, Coward PY, Crook M, Barclay GR, Wilson RF: The short-term effects of treatment of chronic periodontitis on circulating levels of endotoxin, C-reactive protein, tumor necrosis factor-alpha, and interleukin-6. J Periodontol 2004;75:420–428.

13 Willetts IE, Kite P, Barclay GR, Banks RE, Rumley A, Allgar V, Stringer MD: Endotoxin, cytokines and lipid peroxides in children with intussusception. Br J Surg 2001;88:878–83.

14 Clements WD, Erwin P, McCaigue MD, Halliday I, Barclay GR, Rowlands BJ: Conclusive evidence of endotoxaemia in biliary obstruction. Gut 1998;42:293–299.

15 Bayley JP, Ottenhoff THM, Verweij CL: Is there a future for TNF promoter polymorphisms? Genes Immun 2004;5:315–329.

16 Erridge C, Stewart J, Poxton IR: Monocytes heterozygous for the Asp299Gly and Thr399Ile mutations in the Toll-like receptor 4 gene show no deficit in LPS signalling. J Exp Med 2003; 197:1787–1791.

17 Bennett-Guerrero E, Ayuso L, Hamilton-Davies C, White WD, Barclay GR, Smith PK, King SA, Muhlbaier LH, Newman MF, Mythen MG: Relationship of preoperative antiendotoxin core antibodies and adverse outcomes following cardiac surgery. JAMA 1997;277: 646–650.

18 Bennett-Guerrero E, Panah MH, Barclay GR, Bodian CA, Winfree WJ, Andres LA, Reich DL, Mythen MG: Decreased endotoxin immunity is associated with greater mortality and/or prolonged hospitalization after surgery. Anesthesiology 2001;94:992–998.

19 Bennett-Guerrero E, Barclay GR, Youssef ME, Hossain S, Vela-Cantos F, Andres LA, Poxton IR: Exposure to *Bacteroides fragilis* endotoxin during cardiac surgery. Anesth Analg 2000;90: 819–823.

20 Bennett-Guerrero E, McIntosh TJ, Barclay GR, Snyder S, Gibbs RJ, Mythen MG, Poxton IR: Preparation and preclinical evaluation of a novel liposomal complete-core lipopolysaccharide vaccine. Infect Immun 2000;68:6202–6208.

21 Erridge C, Stewart J, Bennett-Guerrero E, McIntosh TJ, Poxton IR: The biological activity of a liposomal complete core lipopolysaccharide vaccine. J Endotoxin Res 2002;8:39–46.

22 Levi K, Towner KJ: Detection of methicillin-resistant *Staphylococcus aureus* (MRSA) in blood with the EVIGENE MRSA detection kit. J Clin Microbiol 2003;41:3890–3892.

23 Rotimi, VO, Khoursheed M, Brazier JS, Jamal WY: *Bacteroides* species highly resistant to metronidazole: An emerging problem. Clin Microbiol Infect 1999;5:166–169.

24 Stubbs SL, Brazier JS, Talbot PR, Duerden BI: PCR-restriction fragment length polymorphism analysis for identification of *Bacteroides* spp and characterization of nitroimidazole resistance genes. J Clin Microbiol 2000;38: 3209–3213.

Review

Medical Principles and Practice

Med Princ Pract 2005;14(suppl 1):27–34
DOI: 10.1159/000086182

Received: April 19, 2004
Revised: July 17, 2004

Mycobacterial Gene Cloning and Expression, Comparative Genomics, Bioinformatics and Proteomics in Relation to the Development of New Vaccines and Diagnostic Reagents

A.S. Mustafa

Department of Microbiology, Faculty of Medicine, Kuwait University, Kuwait

Key Words
Mycobacterium tuberculosis · Antigens · Vaccine

Abstract
Recent advances in molecular and genomic techniques have facilitated research on several aspects of mycobacteriology, such as diagnosis and the identification of new vaccines and therapeutic targets for various diseases, including tuberculosis. The aim of this review was to analyze the implications of advances in molecular and genomic techniques on the development of new vaccines for tuberculosis as well as immunological reagents to diagnose the disease. Gene cloning and expression, DNA and protein sequencing, polymerase chain reaction, comparative genomics, bioinformatics, proteomics and DNA and peptide synthesis coupled with the application of cellular immunology techniques have led to the identification of several antigens of *Mycobacterium tuberculosis,* which have potential for diagnosis and vaccine applications. For example, cross-reactive mycobacterial antigens like heat shock proteins, MTB32 and MTB39, have been identified as new vaccine candidates, and antigens encoded by *M. tuberculosis*-specific genomic regions as new reagents for diagnosis.

Introduction

Tuberculosis (TB) and leprosy are two major mycobacterial diseases that have afflicted humans throughout recorded history. However, due to multidrug therapy and active case finding efforts coordinated through the World Health Organization, impressive advances have been made towards the control of leprosy. For example, in 1970 there were 10.8 million people suffering from leprosy, but the number was reduced to less than 1 million in 1996 and further dropped to about 0.5 million by 2003, and the disease has been eliminated from at least 98 countries [1]. On the other hand, the global problem of TB, instead of improving, has deteriorated due to several factors, including the emergence and increase in the number of patients infected with multidrug-resistant strains of *Mycobacterium tuberculosis,* and TB-HIV coinfection. Currently, about 8 million people develop active disease and 2 million die of TB each year [2]. TB ranks among the top 10 causes of human mortality in the world and the World Health Organization has declared TB 'a global emergency'.

To control the worldwide problem of TB, universally efficacious anti-TB vaccines and reagents for specific diagnosis are urgently needed. With respect to anti-TB vaccines, vaccination with the attenuated *Mycobacterium bovis* Bacillus Calmette-Guérin (BCG) has been widely used for several decades, but its efficacy to protect against the most infectious form of TB (pulmonary disease) in

KARGER

Fax +41 61 306 12 34
E-Mail karger@karger.ch
www.karger.com

1011–7571/05/0147–0027$22.00/0

Accessible online at:
www.karger.com/mpp

Prof. Abu Salim Mustafa
Department of Microbiology, Faculty of Medicine, Kuwait University
PO Box 24923
13110 Safat (Kuwait)
Tel. +965 5312300, ext. 6505, Fax +965 5332719, E-Mail abusalim@hsc.kuniv.edu.kw

adults varies tremendously in different parts of the world. In particular, BCG has shown poor efficacy in the developing countries of Asia and Africa, where TB is highly endemic and vaccine-induced protection is most urgently needed [3]. In addition, the diagnostic value of the presently used skin test reagent, purified protein derivative (PPD) of *M. tuberculosis,* is low due to the presence of antigens that cross-react with environmental mycobacteria and the vaccine strains of *M. bovis* BCG. Furthermore, BCG being a live vaccine can cause disease by itself in immunocompromised subjects for whom vaccination with BCG is contraindicated, including AIDS patients, who are usually at a very high risk of developing TB [4]. Identification of *M. tuberculosis* antigens useful for the specific diagnosis of TB and the development of new TB vaccines that are protective in all parts of the world is therefore an urgent global need.

Gene Cloning and Expression to Identify Antigenic Proteins of *M. tuberculosis*

The advances in gene cloning and expression of mycobacterial genes in heterologous hosts have greatly facilitated the identification of antigenic proteins of *M. tuberculosis* recognized by antibodies and T cells. The first breakthrough in this field was reported by Bloom et al. in 1985 [5]. They used the λ gt11 phage expression system for high-level expression of mycobacterial DNA as fusion proteins (fused to the structural protein β-galactosidase) in *Escherichia coli.* The fusion protein approach assured efficient translation and transcription of DNA in *E. coli.* Another advantage of the fusion protein approach is the relative stability of the mycobacterial polypeptide. The fusion proteins are usually more resistant to the proteolytic machinery of *E. coli* than the foreign polypeptides alone [5].

The mycobacterial recombinant DNA libraries constructed by Bloom et al. [5] were first screened with antibody probes to identify recombinant phage clones expressing the antigenic proteins of mycobacteria. This resulted in the identification of DNA sequences expressing 12-, 14-, 18-, 19-, 28-, 36-, 65- and 70-kDa proteins [5]. When tested for immunological reactivity with T cell clones established from TB patients and healthy donors vaccinated with killed *Mycobacterium leprae,* the 65-, 19- and 18-kDa proteins were found reactive with human T cells [6–8]. In addition, these antigens were also recognized by polyclonal T cell lines established against whole-cell mycobacteria [9]. However, the 65-kDa antigen was found to be reactive with most T cell lines [9], and T cells responding to it were detected even 8 years after vaccination with a preparation of killed *M. leprae* [10]. These results suggested that the T cells reactive with the 65-kDa antigen were long-lasting memory T cells, and therefore vaccination with the 65-kDa antigen may be useful in providing long-term immunity. Furthermore, the 65-kDa antigen was presented to T cells promiscuously and in association with frequently expressed HLA-DR molecules [11, 12]. Thus, if used as a vaccine, this antigen will not have limitations due to HLA restriction. However, the 65-kDa antigen is not useful in the specific diagnosis of TB because it is a cross-reactive heat shock protein (hsp) shared among different mycobacterial species including the vaccine strains of BCG and environmental mycobacteria [13].

The vaccine potential of the 65-kDa mycobacterial hsp (hsp65) was first reported by Silva and Lowrie [14], who demonstrated that immunization with hsp65 expressed in a mouse macrophage cell line J774 protects mice against challenge with *M. tuberculosis.* The adoptive transfer experiments from immunized to naïve mice further showed that antigen-specific T cells were responsible for providing protection against the growth of *M. tuberculosis* in challenged mice [15]. Vaccination with hsp65-plasmid DNA protects mice against *M. tuberculosis* challenge in a prophylactic mode [16]. In addition, vaccination with hsp65-plasmid DNA has also been shown to be effective in immunotherapy of mice preinfected with *M. tuberculosis,* by shifting Th2 response to the protective Th1 type, and providing a sterilization effect, thus preventing the disease due to reactivation [17].

Another approach to identify antigens of *M. tuberculosis* recognized by T cells was established by us in 1988, in which recombinant DNA libraries were directly screened for T cell reactivity by using mycobacterial antigen-specific T cell clones [18]. In this approach, pools of recombinant antigens were prepared from a large number of recombinant phage clones followed by testing the antigenic pools with a pool of T cell clones established against whole-cell mycobacteria [18]. These studies led to the identification and characterization of 24-kDa lipoprotein antigens of *M. tuberculosis* (Rv2945c, LppX) and *M. leprae* (ML0136, LppX), which are absent in environmental mycobacteria [19, 20]. A synthetic peptide approach identified promiscuous T cell epitopes of LppX [20] and suggested that this antigen could be useful in developing a common vaccine against tuberculosis and leprosy.

By using T cell and serological expression cloning in conjunction with peripheral blood mononuclear cells

(PBMC) from healthy donors, T cell lines derived from mice infected with *M. tuberculosis* and sera from active TB patients, a group of scientists working at Corixa Corporation, USA, have identified 8 different Th1-cell-stimulating proteins with molecular mass ranging from 8.4 to 41 kDa [21–25]. Among these proteins, vaccination with MTB32 and MTB39 as fusion proteins has shown to protect 4 different species of animals, i.e. mice, guinea pigs, rabbits and monkeys. The vaccine is safe and effective in naïve animals as well as in animals preinfected with *M. tuberculosis* [25].

Reverse Genetics to Identify Antigens of *M. tuberculosis*

The techniques of reverse genetics have helped to identify the genes of several *M. tuberculosis* antigens that are actively secreted in the culture filtrate of *M. tuberculosis.* The work to identify the secreted protein antigens of *M. tuberculosis* has received considerable attention because of their possible immunodominance and involvement in protective immunity [26–28]. The evaluation of fractionated proteins of short-term culture filtrate (ST-CF) of *M. tuberculosis* in memory immune mice identified low-molecular-weight (6–10 kDa) proteins as the most potent stimulators of the protective cytokine IFN-γ [29]. In the subsequent studies, the low-molecular-weight fraction of ST-CF was used to raise monoclonal antibodies and purify proteins for amino-terminal sequencing. These studies led to the isolation of the gene for the 6-kDa early secreted antigenic target (ESAT6), which was then expressed as a recombinant protein in *E. coli* and purified [29]. Further testing of human T cells from TB patients with recombinant ESAT6 showed that it was the major stimulator of human T cells in proliferation and IFN-γ assays [30]. In addition, the use of overlapping synthetic peptides covering the sequence of the protein showed that multiple epitopes of ESAT6 were presented to Th1 cells in association with frequently expressed HLA class II molecules [31, 32]. The observation that ESAT6 is expressed in all members of the *M. tuberculosis* complex causing human TB, but not in many other mycobacteria including vaccine strains of *M. bovis* BCG [33], indicates that this antigen is relevant to specific diagnosis of TB. This was confirmed by demonstrating that ESAT6 induced high level of IFN-γ responses in TB patients but not in BCG-vaccinated healthy donors, whereas PPD induced equivalent responses in both groups [34]. Vaccination of mice with ESAT6 has been shown to provide protection against *M. tuberculosis* challenge in the mouse model of TB [35]. In addition, vaccination with a single peptide of ESAT6 (aa 51–70) provided a level of protection that was equivalent to the protection achieved after vaccination with complete ESAT6 or *M. bovis* BCG [35]. These findings demonstrate the potential of ESAT6 and its peptide (aa 51–70) in developing subunit and peptide-based vaccines against TB.

The studies summarized above suggest that ESAT6 could be of value in both specific diagnosis as well as in the development of a new vaccine against TB. However, in order to differentiate between vaccination with ESAT6 and infection with *M. tuberculosis,* a single antigen will not suffice. Therefore, studies have been carried out to identify additional antigens of *M. tuberculosis* with diagnostic and vaccine potential.

The need to identify additional antigens present in the low-molecular-weight fractions of ST-CF is further supported by our recent work aimed to evaluate the immune response of TB patients treated with a multidrug regimen for 6 months [36]. The results showed that, in treated patients, antigen-induced proliferation and IFN-γ responses of PBMC in response to the low-molecular-weight fractions F1 and F2 were severalfold higher than the responses obtained after stimulation of PBMC with ESAT6 [36]. Thus, this suggests that the low-molecular-weight fractions of ST-CF contain additional immunologically important proteins of *M. tuberculosis.*

The experiments carried out to identify the proteins in the low-molecular-weight fraction of ST-CF by raising monoclonal antibodies and screening the λ gt11 recombinant DNA library of *M. tuberculosis* with these antibodies led to the identification of 3 other immunogenic proteins termed as CFP10, TB7.3 and TB10.4 [33]. The sequence analyses demonstrated that TB10.4 and CFP10 belong to the ESAT6 family of low-mass proteins, whereas TB7.3 is a low-molecular-mass protein outside this family. TB10.4 and CFP10 have almost identical size (10 kDa) and genes coding for both have approximately 40% identity to *esat6* [33]. The proteins CFP10, TB7.3 and TB10.4 were expressed as recombinant proteins in *E. coli,* and the immunological reactivity of the recombinant proteins was assessed with PBMC from TB patients and *M. bovis* BCG-vaccinated and nonvaccinated healthy donors. In TB patients, the recombinant TB10.4 and CFP10 were strongly recognized and induced IFN-γ release at the same level (CFP10) as or at an even higher level (TB10.4) than ESAT6 [33], whereas TB10.4 showed strong recognition in *M. bovis* BCG-vaccinated healthy subjects as well and TB7.3 induced low level of responses

Table 1. Description of *M. tuberculosis*-specific ORFs predicted in RD1

ORF	Rv gene designation	Encoded protein	Protein length, number of aa
ORF2	*Rv3871*	unknown	591
ORF3	*Rv3872*	PE35	98
ORF4	unknown	unknown	239
ORF5	*Rv3873*	PPE68	371
ORF6	*Rv3874*	CFP10	100
ORF7	*Rv3875*	ESAT6	95
ORF8	unknown	unknown	139
ORF9	*Rv3876*	unknown	666
ORF10	*Rv3877*	unknown	552
ORF11	*Rv3878*	unknown	280
ORF14	unknown	unknown	262
ORF15	unknown	unknown	95

in all the subjects [33]. Thus, CFP10 was found to distinguish TB patients from *M. bovis* BCG-vaccinated donors and could therefore be, together with ESAT6, an interesting candidate for the specific diagnosis of active TB. A comparative evaluation of ESAT6 and CFP10 responses in an enzyme-linked immunospot IFN-γ assay with tuberculin skin testing has revealed that the enzyme-linked immunospot IFN-γ assay with ESAT6 and CFP10 is more accurate than tuberculin skin testing for identification of individuals who have latent TB [37]. Therefore, in addition to the diagnosis of active TB, the use of ESAT6 and CFP10 could improve TB control by more precise targeting of preventive treatment in people latently infected with *M. tuberculosis.*

The studies conducted to determine the distribution of *tb7.3, tb10.4* and *cfp10* genes in *M. tuberculosis, M. bovis* BCG and environmental mycobacteria have shown that all of these genes are present in *M. tuberculosis* but, like *esat6, cfp10* is absent in BCG and most environmental mycobacteria [33]. In addition to specific diagnosis, these antigens could also be useful in the development of new vaccines against TB due to lack of interference from most environmental mycobacteria. Further studies have shown that ESAT6 and CFP10 are expressed from a single operon and genes encoding these low-molecular-weight proteins are transcribed together in *M. tuberculosis* [38]. The 9.5-kb genomic region of *M. tuberculosis* encoding ESAT6 and CFP10 has been termed as RD1 [38]. The analysis of RD1 for open reading frames (ORFs) has shown that, in addition to ESAT6 (ORF7, Rv3875) and CFP10 (ORF6, Rv3874), this genomic region can potentially encode 12 other *M. tuberculosis*-specific proteins from an equal number of ORFs (table 1) [39]. As described below, further studies were conducted to identify other major antigens encoded by RD1 ORFs using recombinant DNA and overlapping synthetic peptide approaches.

Identification of Major *M. tuberculosis* Antigens Encoded by RD1 Using Recombinant DNA Approach

To immunologically characterize the putative proteins encoded by RD1 ORFs, we first attempted to clone and express the putative proteins encoded by RD1 as fusion proteins in *E. coli* using appropriate plasmids as cloning and expression vectors [39, 40]. Subsequently, the recombinant proteins were purified and isolated free of the fusion partner (glutathione *S*-transferase of *Schistosoma japonicum*) by affinity purification on glutathione Sepharose and/or Ni-NTA-agarose affinity matrix and cleavage of the purified fusion proteins by treatment with thrombin protease [39, 40]. However, these procedures could result in the expression and purification of only 6 of the targeted 14 proteins, i.e. ORF3 (PE35, Rv3872), ORF5 (PPE68, Rv3873), ORF6 (CFP10, Rv3874), ORF7 (ESAT6, Rv3875), ORF11 (Rv3878) and ORF14 [39, 40]. The problems with other ORFs included low level of expression, degradation of the mycobacterial proteins and the presence of contaminating *E. coli* components in purified preparations. When tested for antibody reactivity using sera from different groups of donors, the purified ORF14 protein was frequently recognized by antibodies present in the sera of TB patients, but not healthy *M. bovis* BCG-vaccinated donors [39, 41], whereas other proteins showed either low frequency of recognition or were recognized by sera from TB patients as well as healthy donors [39–41]. These results suggest that ORF14 protein could be useful in the serodiagnosis of TB. However, when tested with PBMC from TB patients and healthy donors for Th1 cell responses, ORF14 protein showed weak reactivity in antigen-induced proliferation and IFN-γ assays [42], thus suggesting that the ORF14 protein is not important as a new vaccine candidate against TB.

Overlapping Synthetic Peptides to Identify Antigens Encoded by RD1 Genes

To overcome the problems associated with the recombinant DNA approach in expression and purification of RD1-encoded proteins and to identify additional antigens relevant for diagnosis and vaccine development against TB, we have used pools of overlapping synthetic peptides covering the sequence of each protein and tested these pools for T cell reactivity with PBMC from TB patients and BCG-vaccinated healthy donors in antigen-induced proliferation and IFN-γ assays [43]. The results showed that the RD1 pool was as good in stimulating Th1 cells from both donor groups as complex antigens of *M. tuberculosis,* i.e. whole cells and culture filtrate antigens and whole-cell *M. bovis* BCG [44]. The studies further showed that ORF5 (PPE68, Rv3873), ORF6 (CFP10, Rv3874) and ORF7 (ESAT6, Rv3875) are the major antigens recognized by Th1 cells from TB patients, whereas only ORF5 was the major antigen recognized by Th1 cells from *M. bovis* BCG-vaccinated healthy donors [44]. Similar experiments in the cattle model of TB have shown that PPE68, CFP10 and ESAT6 are also the major stimulators of Th1 cell reactivity in *M.-bovis*-infected but not in noninfected or BCG-vaccinated animals [45].

Comparative Genomics and Bioinformatics to Identify *M. tuberculosis*-Specific Genomic Regions

The availability of the complete genome sequence of *M. tuberculosis* in 1998 [46] has opened up new frontiers of research in TB, including the identification of antigens useful as new vaccine candidates and diagnostic reagents. The comparative genomics and bioinformatics have shown that in addition to RD1, 10 other genomic regions present in *M. tuberculosis* H37Rv (RD4–RD13) are deleted in all strains of BCG [47–49]. These regions cover 108 ORFs of *M. tuberculosis* H37Rv and are deleted from all 13 BCG substrains currently used as anti-TB vaccines in different parts of the world. By using an overlapping synthetic peptide approach, we are in the process of screening these regions to identify additional *M. tuberculosis*-specific antigens useful for diagnosis and vaccine applications [50].

In silico comparisons between *M. tuberculosis* and *M. leprae* genome sequences have revealed that among the 1,439 genes common to both mycobacteria, 219 genes code for proteins that show no similarity with proteins from other organisms [51]. Macroarray and bioinformatic analyses revealed that among the 219 'core' genes conserved among *M. tuberculosis* and *M. leprae,* only few were not present in other mycobacteria, i.e. *Mycobacterium avium, Mycobacterium marinum* and *Mycobacterium smegmatis.* Interestingly, some of these missing genes code for proteins of the ESAT6 family [51].

Diagnostic and Vaccine Potentials of RD1 and Other RD Regions

A number of recent studies demonstrate the diagnostic and vaccine potential of proteins encoded by RD1 and other RD regions. The study of Liu et al. [52] using synthetic peptides covering the sequence of 4 gene products encoded by RD2 (Rv1989c) and RD1 (Rv3873, Rv3878, and Rv3879c) showed that 45 and 53% of the TB patients responded to Rv3879c and Rv3873, respectively, identifying these proteins as major *M. tuberculosis* T cell antigens in humans, while 35 and 25% of the patients responded to Rv3878 and Rv1989c, respectively. Of the 38 BCG-vaccinated donors, 2.6% responded to peptides from Rv3878 and Rv3879c, 7.9% responded to Rv3873, and none responded to Rv1989c [52]. Exclusion of cross-reactive peptides encoded in conserved motifs of Rv3873 increased its specificity to 97.4%. The high specificity of Rv3879c peptides and nonconserved Rv3873 sequences, together with their moderate sensitivity in TB patients, identifies these peptides as candidates for inclusion in new T-cell-based tests for *M. tuberculosis* infection [52].

Using pools of synthetic peptides, Demangel et al. [53] have examined the potential immunogenicity of 4 RD1 products, i.e. Rv3872, Rv3873, Rv3878, and Rv3879c. Rv3873 was the only one to elicit IFN-γ-producing cells in C57BL/6 mice infected with *M. tuberculosis.* Anti-Rv3873 T cells were predominantly raised against an epitope mapped in the N-terminal end of the protein. Their results further show that Rv3873 is an immunogenic product of the RD1 region, which does not interfere with the secretion and immunogenicity of CFP10 and ESAT6. When tested as a His-tagged recombinant protein, Rv3873 stimulated high levels of IFN-γ secretion in peripheral blood mononuclear cells isolated from TB patients, as well as from healthy tuberculin PPD-positive subjects and *M. bovis* BCG-vaccinated donors [54]. Epitope mapping performed with overlapping peptides revealed a broad pattern of T cell recognition comprising both TB-specific epitopes and epitopes also recognized by BCG-vaccinated donors [54]. The immunological evaluation of Rv3878 by Agger et al. [55] showed a prominent T cell response to this antigen in TB patients, but very limited responses were seen in BCG-vaccinated donors.

A study conducted in the cattle model of TB used synthetic peptides covering 13 ORFs from the RD1, RD2, and RD14 regions of the *M. tuberculosis* genome [56]. Pools of overlapping peptides spanning these ORFs were tested in *M. bovis*-infected, BCG-vaccinated and unvaccinated cattle. All the peptide pools were recognized in infected cattle, with responder frequencies varying between 16 and 86%. In particular, 8 highly immunogenic antigens were identified whose potentials as diagnostic reagents or as subunit vaccines warrant further study (Rv1983, Rv1986, Rv3872, Rv3873, Rv3878, Rv3879c, Rv1979c, and Rv1769) [56].

For functional characterization of genes within the RD1 segment, *M. tuberculosis* H37Rv deleted for RD1 (H37Rv:δRD1) was assessed for virulence after infections of the human macrophage-like cell line THP-1, human peripheral blood monocyte-derived macrophages, and C57BL/6 mice. In each of these systems, the H37Rv:δRD1 strain was strikingly less virulent than wild-type *M. tuberculosis* H37Rv and was very similar to BCG controls [57]. These results led to the conclusion that genes within or controlled by RD1 are essential for *M. tuberculosis* virulence and that loss of RD1 was important in BCG attenuation [57]. In addition to *M. bovis* BCG, loss of RD1 has also been suggested to contribute to the attenuation of the other live TB vaccine, i.e. *Mycobacterium microti* [58]. To increase the immunogenicity and protective capacity of these vaccines, both were complemented with different constructs containing the genes encoding ESAT6 and CFP10, as well as a variable number of flanking genes. Only reintroduction of the complete RD1 locus led to full secretion of the antigens and resulted in specific ESAT6-dependent immune responses; this suggests that the flanking genes encode a secretory apparatus [59]. Mice and guinea pigs vaccinated with the recombinant strain BCG: RD1-2F9 were better protected against challenge with *M. tuberculosis,* showing less severe pathology and reduced dissemination of the pathogen, as compared with control animals immunized with BCG alone [59].

Proteomics to Identify *M. tuberculosis* Antigens

The availability of the complete genome sequence of *M. tuberculosis* has revolutionized many areas of tuberculosis research and facilitated functional genomics studies, including the identification of antigens useful in diagnosis and vaccine applications. Proteomics has been widely applied to develop two-dimensional polyacrylamide gel electrophoresis maps and databases, evaluate gene expression profiles under different environmental conditions, assess global changes associated with specific mutations, and define drug targets of bacterial pathogens. Genomics has revealed the sequence of about 4,000 genes in *M. tuberculosis* [46]. Proteomics complements genomics in showing which genes are really expressed and could be useful as drug targets and new vaccine candidates [60]. Moreover, proteomics may identify the expression of genes not predicted by genomics [61].

Classical proteomics combines two-dimensional gel electrophoresis (2-DE) for the separation and quantification of proteins in a complex mixture with mass spectrometric identification of selected proteins. Using these techniques, Jungblut et al. [61] demonstrated the expression of 6 genes of *M. tuberculosis* not predicted by genomics. When coupled to immunological assays, proteomics may also be used to identify B cell and T cell antigens within complex protein mixtures [62]. In a study conducted by Covert et al. [63], large quantities of subcellular protein fractions of *M. tuberculosis* were resolved by two-dimensional liquid-phase electrophoresis, resulting in 355 and 299 fractions of culture filtrate and cytosolic proteins, respectively. Analysis of these fractions against splenocytes of C57Bl/6 mice infected with *M. tuberculosis* resulted in the identification of 37 fractions that stimulated a dominant T cell response, as measured by the production of IFN-γ. Molecular characterization of the proteins contained in each fraction by liquid chromatography-mass spectrometry and liquid chromatography-tandem mass spectrometry resulted in the identification of 30 individual proteins. Although many of these represented previously defined antigens, 17 of these proteins were novel T cell antigens [63]. These data demonstrate that proteomics offers a rapid and facile approach for elucidation of immunodominant T cell antigens of *M. tuberculosis.*

A comprehensive proteomics analysis of culture supernatant (CSN) proteins of *M. tuberculosis* H37Rv was accomplished by combination of 2-DE, mass spectrometry, and N-terminal sequencing by Edman degradation [64]. Analytical 2-DE gels resolved approximately 1,250 protein spots from CSN of *M. tuberculosis* H37Rv, 381 of which were identified by mass spectrometry and/or Edman degradation. The study revealed 137 different proteins, 42 of which had previously been described as secreted [64]. Comparative proteome analysis of CSN from virulent *M. tuberculosis* H37Rv and attenuated *M. bovis* BCG (Copenhagen) identified 39 *M. tuberculosis*-specific spots containing 27 different proteins. These included 5 proteins encoded by ORFs absent from *M. bovis* BCG, e.g. ESAT6, as well as 22 novel differential proteins, such as acetyl-CoA C-acetyltransferase (Rv0243) and 2 putative Esat6-like proteins (Rv1198, Rv1793) [64]. These proteins could provide additional antigens useful in the specific diagnosis and/or development of new vaccines against TB.

Conclusions

The advances in molecular and genomic techniques have led to the identification of several antigens of *M. tuberculosis* including hsp, low-molecular-weight proteins actively secreted in the culture filtrate of *M. tuberculosis* and proteins encoded by regions deleted in BCG and other mycobacteria. However, the antigens belonging to the ESAT6 family of low-molecular-weight proteins appear to be more useful for vaccine and diagnostic applications because of their immunodominance for recognition by protective Th1 cells and specificity for *M. tuberculosis.*

Acknowledgment

The study was supported by Kuwait University Research Administration grant No. MI114.

References

1 Elimination of leprosy as a public health problem. http://www.who.int/lep/.
2 Frieden TR, Sterling TR, Munsiff SS, Watt CJ, Dye C: Tuberculosis. Lancet 2003;362:887–899.
3 Fine PE: Variation in protection by BCG: Implications of and for heterologous immunity. Lancet 1995;346:1339–1345.
4 Schlossberg D: Tuberculosis and nontuberculous mycobacterial infections, ed 4. Philadelphia, WB Saunders Company, 1999, p 3.
5 Bloom BR, Mehra V, Young RA: Genes for the protein antigens of the tuberculosis and leprosy bacilli. Biosci Rep 1985;5:839–845.
6 Mustafa AS, Gill HK, Nerland A, Britton WJ, Mehra V, Bloom BR, Young RA, Godal T: Human T-cell clones recognize a major *M. leprae* protein antigen expressed in *E. coli.* Nature 1986;319:63–66.
7 Mustafa AS, Oftung F, Gill HK, Natvig I: Characteristics of human T-cell clones from BCG and killed *M. leprae* vaccinated subjects and tuberculosis patients. Recognition of recombinant mycobacterial antigens. Lepr Rev 1986;57(suppl 2):123–130.
8 Oftung F, Mustafa AS, Husson R, Young RA, Godal T: Human T cell clones recognize two abundant *Mycobacterium tuberculosis* protein antigens expressed in *Escherichia coli.* J Immunol 1987;138:927–931.
9 Mustafa AS: Identification of T-cell-activating recombinant antigens shared among three candidate antileprosy vaccines, killed *M. leprae, M. bovis* BCG, and mycobacterium w. Int J Lepr Other Mycobact Dis 1988;56:265–273.
10 Mustafa AS, Oftung F: Long-lasting T-cell reactivity to *Mycobacterium leprae* antigens in human volunteers vaccinated with killed *M. leprae.* Vaccine 1993;11:1108–1112.
11 Mustafa AS, Lundin KE, Oftung F: Human T cells recognize mycobacterial heat shock proteins in the context of multiple HLA-DR molecules: Studies with healthy subjects vaccinated with *Mycobacterium bovis* BCG and *Mycobacterium leprae.* Infect Immun 1993;61:5294–5301.
12 Mustafa AS, Lundin KE, Meloen RH, Shinnick TM, Oftung F: Identification of promiscuous epitopes from the mycobacterial 65-kilodalton heat shock protein recognized by human CD4(+) T cells of the *Mycobacterium leprae* memory repertoire. Infect Immun 1999;67:5683–5689.
13 Mustafa AS: HLA-restricted immune response to mycobacterial antigens: Relevance to vaccine design. Hum Immunol 2000;61:166–171.
14 Silva CL, Lowrie DB: A single mycobacterial protein (hsp 65) expressed by a transgenic antigen-presenting cell vaccinates mice against tuberculosis. Immunology 1994;82:244–248.
15 Silva CL, Silva MF, Pietro RC, Lowrie DB: Protection against tuberculosis by passive transfer with T-cell clones recognizing mycobacterial heat-shock protein 65. Immunology 1994;83:341–346.
16 Tascon RE, Colston MJ, Ragno S, Stavropoulos E, Gregory D, Lowrie DB: Vaccination against tuberculosis by DNA injection. Nat Med 1996;2:888–892.
17 Lowrie DB, Tascon RE, Bonato VL, Lima VM, Faccioli LH, Stavropoulos E, Colston MJ, Hewinson RG, Moelling K, Silva CL: Therapy of tuberculosis in mice by DNA vaccination. Nature 1999;400:269–271.
18 Mustafa AS, Oftung F, Deggerdal A, Gill HK, Young RA, Godal T: Gene isolation with human T lymphocyte probes. Isolation of a gene that expresses an epitope recognized by T cells specific for *Mycobacterium bovis* BCG and pathogenic mycobacteria. J Immunol 1988;141:2729–2733.
19 Oftung F, Wiker HG, Deggerdal A, Mustafa AS: A novel mycobacterial antigen relevant to cellular immunity belongs to a family of secreted lipoproteins. Scand J Immunol 1997;46:445–451.
20 Al-Attiyah R, Mustafa AS: Computer-assisted prediction of HLA-DR binding and experimental analysis for human promiscuous Th1-cell peptides in the 24 kDa secreted lipoprotein (LppX) of *Mycobacterium tuberculosis.* Scand J Immunol 2004;59:16–24.
21 Coler RN, Skeiky YA, Vedvick T, Bement T, Ovendale P, Campos-Neto A, Alderson MR, Reed SG: Molecular cloning and immunologic reactivity of a novel low molecular mass antigen of *Mycobacterium tuberculosis.* J Immunol 1998;16:2356–2364.
22 Dillon DC, Alderson MR, Day CH, Lewinsohn DM, Coler R, Bement T, Campos-Neto A, Skeiky YA, Orme IM, Roberts A, Steen S, Dalemans W, Badaro R, Reed SG: Molecular characterization and human T-cell responses to a member of a novel *Mycobacterium tuberculosis* mtb39 gene family. Infect Immun 1999;67:2941–2950.
23 Skeiky YA, Lodes MJ, Guderian JA, Mohamath R, Bement T, Alderson MR, Reed SG: Cloning, expression, and immunological evaluation of two putative secreted serine protease antigens of *Mycobacterium tuberculosis.* Infect Immun 1999;67:3998–4007.
24 Alderson MR, Bement T, Day CH, Zhu L, Molesh D, Skeiky YA, Coler R, Lewinsohn DM, Reed SG, Dillon DC: Expression cloning of an immunodominant family of *Mycobacterium tuberculosis* antigens using human CD4(+) T cells. J Exp Med 2000;191:551–560.
25 Reed S: TB vaccine development: Preclinical activities. Abstract session 5. Novel vaccines/vaccine strategies for TB. First Int Conf on TB Vaccines for the World, Montreal, September 2003.
26 Andersen P: Host responses and antigens involved in protective immunity to *Mycobacterium tuberculosis.* Scand J Immunol 1997;45:115–131.
27 Mustafa AS, Shaban FA, Abal AT, Al-Attiyah R, Wiker HG, Lundin KE, Oftung F, Huygen K: Identification and HLA restriction of naturally derived Th1-cell epitopes from the secreted *Mycobacterium tuberculosis* antigen 85B recognized by antigen-specific human CD4(+) T-cell lines. Infect Immun 2000;68:3933–3940.

28 Al-Attiyah R, Shaban FA, Wiker HG, Oftung F, Mustafa AS: Synthetic peptides identify promiscuous human Th1 cell epitopes of the secreted mycobacterial antigen MPB70. Infect Immun 2003;71:1953–960.
29 Andersen P, Andersen AB, Sorensen AL, Nagai S: Recall of long-lived immunity to *Mycobacterium tuberculosis* infection in mice. J Immunol 1995;154:3359–3372.
30 Mustafa AS, Amoudy HA, Wiker HG, Abal AT, Ravn P, Oftung F, Andersen P: Comparison of antigen-specific T-cell responses of tuberculosis patients using complex or single antigens of *Mycobacterium tuberculosis.* Scand J Immunol 1998;48:535–543.
31 Mustafa AS, Oftung F, Amoudy HA, Madi NM, Abal AT, Shaban F, Rosen Krands I, Andersen P: Multiple epitopes from the *Mycobacterium tuberculosis* ESAT-6 antigen are recognized by antigen-specific human T cell lines. Clin Infect Dis 2000;30(suppl 3):S201–S205.
32 Mustafa AS, Shaban FA, Al-Attiyah R, Abal AT, El-Shamy AM, Andersen P, Oftung F: Human Th1 cell lines recognize the *Mycobacterium tuberculosis* ESAT-6 antigen and its peptides in association with frequently expressed HLA class II molecules. Scand J Immunol 2003;57:125–134.
33 Skjot RL, Oettinger T, Rosenkrands I, Ravn P, Brock I, Jacobsen S, Andersen P: Comparative evaluation of low-molecular-mass proteins from *Mycobacterium tuberculosis* identifies members of the ESAT-6 family as immunodominant T-cell antigens. Infect Immun 2000; 68:214–220.
34 Ravn P, Demissie A, Eguale T, Wondwosson H, Lein D, Amoudy HA, Mustafa AS, Jensen AK, Holm A, Rosenkrands I, Oftung F, Olobo J, von Reyn F, Andersen P: Human T cell responses to the ESAT-6 antigen from *Mycobacterium tuberculosis.* J Infect Dis 1999;179: 637–645.
35 Olsen AW, Hansen PR, Holm A, Andersen P: Efficient protection against *Mycobacterium tuberculosis* by vaccination with a single subdominant epitope from the ESAT-6 antigen. Eur J Immunol 2000;30:1724–1732.
36 Al-Attiyah R, Mustafa AS, Abal AT, Madi NM, Andersen P: Restoration of mycobacterial antigen-induced proliferation and interferon-gamma responses in peripheral blood mononuclear cells of tuberculosis patients upon effective chemotherapy. FEMS Immunol Med Microbiol 2003;38:249–256.
37 Ewer K, Deeks J, Alvarez L, Bryant G, Waller S, Andersen P, Monk P, Lalvani A: Comparison of T-cell-based assay with tuberculin skin test for diagnosis of *Mycobacterium tuberculosis* infection in a school tuberculosis outbreak. Lancet 2003;361:1168–1173.
38 Berthet FX, Rasmussen PB, Rosenkrands I, Andersen P, Gicquel B: A *Mycobacterium tuberculosis* operon encoding ESAT-6 and a novel low-molecular-mass culture filtrate protein (CFP-10). Microbiology 1998;144:3195–3203.
39 Ahmad S, Amoudy HA, Thole JE, Young DB, Mustafa AS: Identification of a novel protein antigen encoded by a *Mycobacterium tuberculosis*-specific RD1 region gene. Scand J Immunol 1999;49:515–522.
40 Ahmad S, Ali MM, Mustafa AS: Construction of a modified vector for efficient purification of recombinant *Mycobacterium tuberculosis* proteins expressed in *Escherichia coli.* Protein Expr Purif 2003;29:167–175.
41 Mustafa AS, Al-Attiyah R: Tuberculosis: Looking beyond BCG vaccines. J Postgrad Med 2003;49:134–140.
42 Mustafa AS: Biotechnology in the development of new vaccines and diagnostic reagents against tuberculosis. Curr Pharm Biotechnol 2001;2:157–173.
43 Mustafa AS: Development of new vaccines and diagnostic reagents against tuberculosis. Mol Immunol 2002;39:113–119.
44 Mustafa AS: Progress towards the development of new anti-tuberculosis vaccines; in: Progress in Tuberculosis Research. Kuwait. New York, Nova Science Publishers, Inc., Editor: LC Smithe, 2005, p 47–76.
45 Mustafa AS, Cockle PJ, Shaban F, Hewinson RG, Vordermeier HM: Immunogenicity of *Mycobacterium tuberculosis* RD1 region gene products in infected cattle. Clin Exp Immunol 2002;130:37–42.
46 Cole ST, Brosch R, Parkhill J, Garnier T, Churcher C, Harris D, et al: Deciphering the biology of *Mycobacterium tuberculosis* from the complete genome sequence. Nature 1998; 393:537–544.
47 Behr MA, Wilson MA, Gill WP, Salamon H, Schoolnik GK, Rane S, Small PM: Comparative genomics of BCG vaccines by whole-genome DNA microarray. Science 1999;284: 1520–1523.
48 Gordon SV, Brosch R, Billault A, Garnier T, Eiglmeier K, Cole ST: Identification of variable regions in the genomes of tubercle bacilli using bacterial artificial chromosome arrays. Mol Microbiol 1999;32:643–655.
49 Brosch R, Gordon SV, Pym A, Eiglmeier K, Garnier T, Cole ST: Comparative genomics of the mycobacteria. Int J Med Microbiol 2000; 290:143–152.
50 Mustafa AS, Al-Attiyah R: *Mycobacterium tuberculosis* antigens and peptides as new vaccine candidates and immunodiagnostic reagents against tuberculosis. Kuwait Med J 2004;36:171–176.
51 Marmiesse M, Brodin P, Buchrieser C, Gutierrez C, Simoes N, Vincent V, Glaser P, Cole ST, Brosch R: Macro-array and bioinformatic analyses reveal mycobacterial ‘core’ genes, variation in the ESAT-6 gene family and new phylogenetic markers for the *Mycobacterium tuberculosis* complex. Microbiology 2004;150: 483–496.
52 Liu XQ, Dosanjh D, Varia H, Ewer K, Cockle P, Pasvol G, Lalvani A: Evaluation of T-cell responses to novel RD1- and RD2-encoded *Mycobacterium tuberculosis* gene products for specific detection of human tuberculosis infection. Infect Immun 2004;72:2574–2581.
53 Demangel C, Brodin P, Cockle PJ, Brosch R, Majlessi L, Leclerc C, Cole ST: Cell envelope protein PPE68 contributes to *Mycobacterium tuberculosis* RD1 immunogenicity independently of a 10-kilodalton culture filtrate protein and ESAT-6. Infect Immun 2004;72:2170–2176.
54 Okkels LM, Brock I, Follmann F, Agger EM, Arend SM, Ottenhoff TH, Oftung F, Rosenkrands I, Andersen P: PPE protein (Rv3873) from DNA segment RD1 of *Mycobacterium tuberculosis:* Strong recognition of both specific T-cell epitopes and epitopes conserved within the PPE family. Infect Immun 2003;71: 6116–6123.
55 Agger EM, Brock I, Okkels LM, Arend SM, Aagaard CS, Weldingh KN, Andersen P: Human T-cell responses to the RD1-encoded protein TB27.4 (Rv3878) from *Mycobacterium tuberculosis.* Immunology 2003;110:507–512.
56 Cockle PJ, Gordon SV, Lalvani A, Buddle BM, Hewinson RG, Vordermeier HM: Identification of novel *Mycobacterium tuberculosis* antigens with potential as diagnostic reagents or subunit vaccine candidates by comparative genomics. Infect Immun 2002;70:6996–7003.
57 Lewis KN, Liao R, Guinn KM, Hickey MJ, Smith S, Behr MA, Sherman DR: Deletion of RD1 from *Mycobacterium tuberculosis* mimics bacille Calmette-Guerin attenuation. J Infect Dis 2003;187:117–123.
58 PymAS, Brodin P, Brosch R, Huerre M, Cole ST: Loss of RD1 contributed to the attenuation of the live tuberculosis vaccines *Mycobacterium bovis* BCG and *Mycobacterium microti.* Mol Microbiol 2002;46:709–717.
59 Pym AS, Brodin P, Majlessi L, Brosch R, Demangel C, Williams A, Griffiths KE, Marchal G, Leclerc C, Cole ST: Recombinant BCG exporting ESAT-6 confers enhanced protection against tuberculosis. Nat Med 2003;9: 533–539.
60 Betts JC: Transcriptomics and proteomics: Tools for the identification of novel drug targets and vaccine candidates for tuberculosis. IUBMB Life 2002;53:239–242.
61 Jungblut PR, Muller EC, Mattow J, Kaufmann SH: Proteomics reveals open reading frames in *Mycobacterium tuberculosis* H37Rv not predicted by genomics. Infect Immun 2001;69: 5905–5907.
62 Dobos KM, Spencer JS, Orme IM, Belisle JT: Proteomic approaches to antigen discovery. Methods Mol Med 2004;94:3–17.
63 Covert BA, Spencer JS, Orme IM, Belisle JT: The application of proteomics in defining the T cell antigens of *Mycobacterium tuberculosis.* Proteomics 2001;1:574–586.
64 Mattow J, Schaible UE, Schmidt F, Hagens K, Siejak F, Brestrich G, Haeselbarth G, Muller EC, Jungblut PR, Kaufmann SH: Comparative proteome analysis of culture supernatant proteins from virulent *Mycobacterium tuberculosis* H37Rv and attenuated *M. bovis* BCG Copenhagen. Electrophoresis 2003;24:3405–3420.

Review

Medical Principles and Practice

Med Princ Pract 2005;14(suppl 1):35–48
DOI: 10.1159/000086183

Received: April 18, 2002
Revised: February 19, 2005

Mechanisms of Drug Resistance in Cancer Chemotherapy

Y.A. Luqmani

Department of Pharmaceutical Chemistry, Faculty of Pharmacy, Kuwait University, Kuwait

Key Words
Cancer · Chemotherapy · Drug resistance · Multidrug resistance · P-Glycoprotein

Abstract
The management of cancer involves procedures, which include surgery, radiotherapy and chemotherapy. Development of chemoresistance is a persistent problem during the treatment of local and disseminated disease. A plethora of cytotoxic drugs that selectively, but not exclusively, target actively proliferating cells include such diverse groups as DNA alkylating agents, antimetabolites, intercalating agents and mitotic inhibitors. Resistance constitutes a lack of response to drug-induced tumour growth inhibition; it may be inherent in a subpopulation of heterogeneous cancer cells or be acquired as a cellular response to drug exposure. Resistance varies. Although regulatory approval may require efficacy in as few as 20% of trial cohorts, a drug may subsequently be used in unselected patients displaying resistance to the treatment. Principal mechanisms may include altered membrane transport involving the P-glycoprotein product of the multidrug resistance (MDR) gene as well as other associated proteins, altered target enzyme (e.g. mutated topoisomerase II), decreased drug activation, increased drug degradation due to altered expression of drug-metabolising enzymes, drug inactivation due to conjugation with increased glutathione, subcellular redistribution, drug interaction, enhanced DNA repair and failure to apoptose as a result of mutated cell cycle proteins such as p53. Attempts to overcome resistance mainly involve the use of combination drug therapy using different classes of drugs with minimally overlapping toxicities to allow maximal dosages and with narrowest cycle intervals, necessary for bone marrow recovery. Adjuvant therapy with P-glycoprotein inhibitors and, in specific instances, the use of growth factor and protein kinase C inhibitors are newer experimental approaches that may also prove effective in abrogating or delaying onset of resistance. Gene knockout using antisense molecules may be another effective way of blocking drug resistance genes. Conversely, drug resistance may also be used to good purpose by transplanting retrovirally transformed CD34 cells expressing the MDR gene to protect the bone marrow during high-dose chemotherapy.

Types of Cancer Treatments

Therapeutic procedures for cancer patients still remain largely empirical and vary between centres. Treatment depends upon a variety of individual factors, which may include the specific pathological and molecular characteristics of the cancer, its location, extent of disease and the health status of the patient.

The ultimate objective is to destroy all the cancer cells whilst inflicting minimal damage on the normal tissue.

KARGER

Fax +41 61 306 12 34
E-Mail karger@karger.ch
www.karger.com

1011–7571/05/0147–0035$22.00/0

Accessible online at:
www.karger.com/mpp

Prof. Y.A. Luqmani
Department of Pharmaceutical Chemistry, Faculty of Pharmacy, Kuwait University
PO Box 24923
13110 Safat (Kuwait)
Tel. +965 533 7135, Fax +965 5342807, E-Mail yunus@hsc.edu.kw

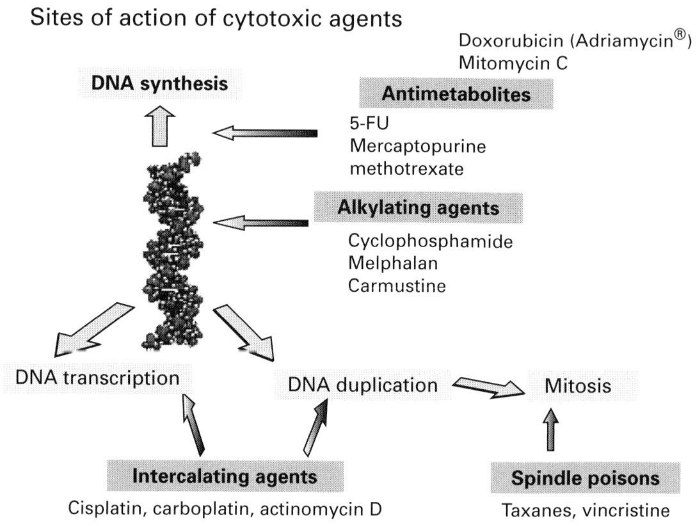

Fig. 1. Sites of action of cytotoxic agents.

This can be achieved in a number of ways, either directly or indirectly by depriving the cancer cells of signals needed for cellular proliferation or by stimulating an immune response.

There are several different types of treatment, which may be used alone or in combination, either simultaneously or sequentially: surgery, radiotherapy and drugs. *Surgery* is most frequently the first line of therapy, and for 'early' cancers, it may be curative. *Radiotherapy* is most often used in a localised setting and in conjunction with surgical procedures. The use of drugs involves *chemotherapy* (CTX), which employs a wide group of drugs that have cytotoxic effects which preferentially, but not exclusively, target the rapidly dividing cancer cells. *Endocrine therapy* is a more specific form of treatment, used for example for breast cancer. It is aimed at preventing cancer cell proliferation by antagonising the oestrogen-stimulated intracellular signals which control growth in cells overexpressing the oestrogen receptor. This may be effective in 60–70% of breast cancer patients. *Antibodies* against specific proteins that are overexpressed in cancer cells may be used to preferentially target those cells. Toxin molecules linked to the antibodies would be endocytosed into the cells and destroy them, such as Herceptin, which targets the erbB2/HER2/neu receptor [1–4]. *Targeted therapies* have been employed exploiting the overexpression in many cancer cells of proteins such as the folate receptor [5, 6]. *Metabolic inhibitors* are being investigated, which target specific proteins and pathways involved in cell growth that operate preferentially in cancer cells. Many of these are associated with cell cycle regulation. *Biological response modifiers* are a class of compounds, which may be modulated to elicit an immune response against cancer cells. *Vaccines* may offer an ideal solution, but so far the lack of tumour-specific proteins has hampered this approach. Of these, chemotherapy and endocrine therapy are currently in use. Others are experimental or hopeful approaches for the future. This review describes the mechanisms of drug resistance in cancer chemotherapy.

Chemotherapy

Although adjuvant CTX is frequently used for primary tumours, its main use is to control overt disseminated disease. The excessively active growth-signaling pathways in cancer cells makes them susceptible to a wide range of drugs which target growth-signaling molecules and/or processes involved in cellular replication and expression (fig. 1). However, as these processes also drive normal cells, the effect is preferential and not exclusive, which results in the unwanted side-effects seen with these agents. Cells which are normally actively dividing, in particular the bone marrow constituents and those of the intestinal lining, are particularly susceptible. Disregulated cell cycle events, due to mutations in cancer cells, do sometimes offer opportunities to target those cells without affecting normal cells.

The relatively wide spectrum of activity of cytotoxic drugs makes them a rather harsh and non-specific form of treatment that can only be tolerated for short periods. Indeed the effects of the treatment may sometimes cause more distress than the disease. These side-effects include dry flaky skin, loss of hair, nausea and vomiting, changes in taste and appetite, blood clotting problems, fatigue, depressed immune system and possible sterility. Most side-effects subside after the treatment is over, but sometimes there is permanent damage to the kidneys, heart, lungs or reproductive system. In general, however, benefits outweigh the disadvantages, and CTX is the commonest form of cancer therapy. This is less a reflection of its effectiveness than the lack of a better alternative. The other chemical therapies in the list above constitute some of these alternatives, but at present they play a subsidiary role in the armory of the oncologist.

Many of the CTX drugs that are employed are naturally occurring compounds extracted from plants, while others are synthetic. They can be divided into three major groups on the basis of their mode and site of action: antimetabolites, genotoxic agents and mitotic spindle inhibitors [7].

Antimetabolites

Examples of such drugs include folate, pyrimidine and purine antagonists.

Folate Antagonists or Antifolates. These are inhibitors of dihydrofolate reductase (DHFR), an enzyme involved in nucleotide metabolism. Folates are co-enzymes required for methylation and necessary for the formation of purines and thymidylate. Methotrexate reversibly binds DHFR after entering the cell via the low pH folate transporter, or by reduced folate carriers, reducing the available DHFR for reduction of the tetrahydrofolate precursors, i.e. folic acid and dihydrofolic acid. Methotrexate is used to treat a variety of malignancies including acute lymphocytic leukaemia, large cell lymphoma, high-grade lymphoma, choriocarcinoma and cancer of the breast, bladder, head and neck and bone as well as many inflammatory diseases.

Pyrimidine Antagonists. Three such compounds that are commonly used in cancer therapy are 5-fluorouracil (5-FU; Adrucil®, Fluorouracil, Efudex®, Fluoroplex®), gemcitabine (Gemzar®) and arabinosylcytosine. These may block pyrimidine nucleotide formation or cause premature termination by themselves being incorporated into newly synthesized DNA. 5-FU also affects the processing of ribosomal, transfer and small nuclear RNAs. It may inhibit cell growth by preventing synthesis of thymine nucleotides from uracil nucleotides. It is used extensively in treatment of cancers of the breast, colon, stomach, rectum and pancreas.

Purine Antagonists. These inhibit synthesis of adenine and guanine, and the main examples are acyclovir, 6-mercaptopurine (Purinethol®, Puri-Nethol®) and 6-thioguanine. They are used to treat particularly acute lymphocytic or myelocytic leukaemia, lymphoblastic leukaemia and acute myelogenous and myelomonocytic leukaemias.

Genotoxic Agents

These either bind to DNA or indirectly damage it by affecting enzymes involved in replication, which leads to induced apoptosis. Hence, proliferating cancer cells are more sensitive than most normal cells, which undergo division less frequently. This class of drugs may be subdivided into several groups.

Alkylating Agents. These modify the bases of DNA, interfering with replication and transcription and leading to mutations. Cross-linking atoms in the DNA, preventing strand separation for synthesis or transcription, also cause damage. Base mispairing between strands is also induced by alkylation. Examples include cisplatin, cyclophosphamide, melphalan, mitomycin C and temozolomide.

Intercalating Agents. These bind in the grooves in the DNA helix, interfering with polymerase activity during replication/transcription. Examples are epirubicin and doxorubicin.

Enzyme Inhibitors. These block replication by inhibiting enzymes, such as topoisomerases. The mutagenic properties of these drugs make them carcinogenic and therefore their use entails additional risk of secondary cancers, such as leukaemia. Examples of these are etoposide, topotecan and irinotecan.

Mitotic Spindle Inhibitors

These disrupt mitosis by affecting the formation/function of spindle microtubule fibres required for chromosome alignment. They prevent the polymerisation of tubulin monomers and act in a cell cycle-dependent manner. They also affect normal cells but to a much lesser extent due to the lower frequency of cell division. The commonly used drugs in this category are the plant-derived vinca alkaloids and paclitaxel (Taxol®) and the synthetic docetaxel (Taxotere®).

Failure of Chemotherapy

In some cases, tumours may be refractory to treatment with some types of cytotoxic drugs. In many, if not most instances, patients who initially respond to CTX invariably show a loss of response later on, resulting in tumour regrowth. There are two probable causes for this: (a) the tumour cells may be inherently resistant possibly due to some genetic characteristic or (b) they may acquire resistance following exposure to the drug. In a typical scenario, 1 in 10^6–10^7 cancer cells in a tumour is likely to have inherent resistance against a particular drug. A clinically detectable tumour has in the region of 10^9 cells and it might therefore be expected to contain 10–1,000 drug-resistant cells, which have the potential to repopulate the tumour despite destruction of the sensitive cells. Considering this, the probability of cure or containment is directly related to tumour size at onset of treatment. Aggressive CTX may often cure certain cancers, particularly childhood leukaemias, Hodgkin's disease or testicular cancers, because the few surviving drug-resistant cells may become susceptible to the immune system. Irrespective of how resistance to CTX arises, it poses a serious impediment to the successful elimination of the entire tumour mass. Drug resistance has been described as the single most common reason for discontinuation of a drug [8].

So why does drug resistance vary between individuals and for different drugs? The answer to this is probably quite simple. In order to obtain approval for use in cancer therapy, a drug may, in the original clinical trials, have been shown to be effective in anything from just 20% upwards of a relatively small and possibly selected patient group. Afterwards it is likely that it is used on patients without any preselection. Attempts to overcome the resistance by the use of a different drug is the most obvious approach, but this offers only temporary respite and eventually results in the phenomenon of multidrug resistance (MDR), which is defined as the insensitivity of various tumours to a variety of chemically related or unrelated anticancer drugs, mediated by a process of inactivating the drug or removing it from the target tumour cells.

Potential Mechanisms of Resistance

Various hypotheses, some with more supporting experimental evidence than others, have been proposed to account for the phenomenon of drug resistance. Ironically, some of the mechanisms that could be utilised by cancer cells to resist cytotoxic drugs are probably evolved in normal cells as a defence mechanism against environmental carcinogens. The hypothesis include altered transport of the drug across the plasma membrane, genetic responses, enhanced DNA repair, alteration in target molecules, access to target cells, metabolic effects and growth factors.

Altered Membrane Transport

Probably one of the most significant forms of resistance against the variety of currently used antineoplastic agents is by the action of a group of membrane proteins which extrude cytotoxic molecules, keeping intracellular drug concentration below a cell-killing threshold. These ATP-dependent multidrug transporters belong to the ubiquitous superfamily of ATP-binding cassette (ABC) proteins, which have also been implicated in the resistance to infectious diseases such as AIDS and malaria. These proteins modulate absorption, distribution and excretion of many pharmacological compounds. There may be as many as 48 genes encoding ABC transporters. The ABC proteins have been grouped into 7 subclasses ranging from ABCA to ABCG [9–11] based upon genomic organization, order of domains and sequence homology.

The MDR1 gene, which encodes P-glycoprotein (P-gp; MDR1, ABCB1) [12–15], a phosphorylated and glycosylated 170-kDa protein of 1,280 amino acids, is the most widely observed mechanism in clinical transport-associated MDR [16–19]. Several other ABC transporters have been described: the MDR-associated protein 1 (MRP1, ABCC1), the mitoxantrone resistance protein (MXR1/BCRP, ABCG2) [19–23] and the ABCB4 (MDR3) and ABCB11 (sister P-gp or BSEP) proteins involved in the secretion of hepatic phosphatidyl choline and bile acids, respectively [24], as well as transport of certain drugs [25]. In addition to MRP1, five homologues (MRP2–MRP6) have been cloned and overexpression of MRP2 was shown to confer MDR [21, 26]. MRP3 and MRP5, which are transporters of organic conjugates and nucleosides, respectively, are also implicated in drug resistance [21]. The MRPs ubiquitously expressed in normal tissues and to varying degrees in human cancers [27] belong to the ABCC subfamily comprising 11 genomic regions. Many of these proteins have been identified as ATP-dependent membrane transporters for various drugs and anions, with some such as MRP1 also utilising glutathione [28, 29].

The most intriguing characteristic distinguishing MDR proteins from other mammalian transporters is their highly promiscuous substrate specificity [30]. In contrast with selective (classical) transport proteins, multidrug transporters, as well as handling unique compounds, translocate a large number of structurally diverse mainly hydrophobic compounds, which explains the cross-resistance to several chemically unrelated compounds, the hallmark of the MDR phenotype [16–19]. Tumours with MDR protein overexpression, such as hepatomas, lung or colon carcinomas, frequently show intrinsic resistance. P-gp transports large hydrophobic, uncharged or slightly positively charged compounds, while the MRP family is the main transport vehicle for hydrophobic anionic conjugates and extrudes uncharged hydrophobic drugs. Table 1 is a list of commonly used drugs, including important anticancer agents, which are extruded by P-gp.

The precise physiological function of P-gp in the absence of therapeutics or toxins is probably a protective role. Results of gene knockout experiments in mice support the idea that MDR proteins play a part in protecting the cell from xenobiotics [31]. Tissue distribution of MDR-ABC proteins is quite varied. P-gp is often difficult to detect because of its low level of expression, but it is ubiquitous in those tissues involved in absorption and secretion [16–19] (e.g. epithelial cells of the kidney proximal tubules, superficial columnar epithelial cells of the colon and jejunum, hepatocytes, epithelial cells of pancreatic small ductules, adrenal medulla and cortex) [32] and in a variety of immune cells [33–36] where it probably contributes significantly to the removal of drugs and toxins from the bone marrow. Its expression varies between individuals and is a function of both genotypic characteristics and of general metabolic/environmental conditions, which may include exposure to heat shock, irradiation, genotoxic stress and inflammatory stimuli as well as cytokines and growth factors. It has been shown that P-gp-mediated drug transport can be stimulated by the antihypertensive prazosin and by progesterone [37].

The promise of preclinical data on the role of P-gp has been somewhat disappointing in terms of the clinical experience with P-gp modulators to overcome drug resistance [38]. Although P-gp appears to play a very important role in the CNS penetration of drugs, its effect on drug absorption may not be as important as generally believed [39]. In some studies, P-gp has been reported to have prognostic significance in certain types of neoplasms [40–42], but very often it has generally failed to show a correlation with clinical response [43–45]. This has led to much controversy regarding the precise role of P-gp in clinically relevant tumour drug resistance. Its precise mode of action also remains a perplexing problem. There have been suggestions that the MDR proteins do not transport drugs per se but alter ion transport or signal transduction, which then affects drug distribution [46].

Table 1. Selected list of drugs that are extruded by P-gp

Cancer drugs	HIV protease inhibitors
Doxorubicin	Amprenavir
Daunorubicin	Indinavir
Vinblastine	Nelfinavir
Vincristine	Ritonavir
Actinomycin D	Saquinavir
Paclitaxel	Cardiac drugs
Teniposide	Digoxin
Etoposide	Quinidine
Immunosuppressive drugs	Anti-emetic
Cyclosporin A	Ondansetron
FK506	Antidiarrhoeal agent
Lipid-lowering agent	Loperamide
Lovastatin	Antigout agent
Antihistamine	Colchicine
Terfenadine	Antibiotic
Steroids	Erythromycin
Aldosterone	Antihelminthic agent
Hydrocortisone	Ivermectin
Cortisol	Antituberculous agent
Corticosterone	Rifampin
Dexamethasone	Fluorescent dye
Dopamine antagonist	Rhodamine-123
Domperidone	

Anticancer drugs and cytotoxic cytokines such as members of the TNF/Fas ligand family play a predominant role in apoptosis induction in tumour cells and are critical in cancer therapy. Given the wide-ranging actions of P-gp, one group has extensively examined its association with cell death [47]. Pathways leading to apoptosis appear to involve a family of cysteine aspases known as caspases [48]. Using drug-resistant tumour cell lines, their results suggest that P-gp confers resistance to Fas-induced caspase-3 activation and apoptosis. P-gp-expressing cells are resistant to a wide range of stimuli that activate the caspase apoptotic cascade, but are not resistant to caspase-independent cell death mediated by pore-forming proteins and GzB. Inhibition of P-gp with antibody inhibitors completely reverses the resistance to caspase-dependent cell death [49, 50]. They also demonstrated ATP-dependent inhibition of Fas-induced caspase-8 activation but not formation of the death-inducing signal complex [51]. Cullen et al. [52], however, found that an MRP-overexpressing MDR leukaemia subline had the same level of Fas expression as the parental cells and similar apoptotic responses to anti-Fas antibody, questioning the hypothesis that selection of drug resistance results in resistance to Fas-mediated apoptosis.

Genetic Responses

Drugs such as methotrexate inhibit key enzymes in pathways controlling proliferation. Thus, increased transcription of the gene that encodes the enzyme can lead to increased levels of the target. As the drug concentration will be limited, it will be unable to block the additional enzyme that is being synthesised, and therefore the cell will effectively overcome the inhibitory effect. One manner in which expression can be elevated is through the process of gene amplification, which involves selective replication of a region of the chromosome, resulting in multiple copies of the same gene, with each one being transcribed to produce a high level of mRNA.

P-gp activity is also related to the status and level of activity of the MDR1 gene. Rifampicin can induce MDR1 expression. Induction of intestinal P-gp by rifampicin has been shown to be the major mechanism responsible for reduced digoxin levels during concomitant rifampicin therapy; in healthy male volunteers, the oral bio-availability of digoxin decreased by 30% and intestinal P-gp levels were induced 3.5-fold during rifampicin therapy [53, 54].

Mickley et al. [55] found that in several drug-resistant cancer cell lines as well as in DNA from two leukaemic patients who had developed drug resistance, gene rearrangements had occurred resulting in initial activation or increased expression of MDR1. Polymorphism in exon 26 (C3435T) of MDR1 is significantly correlated with levels of expression and function of MDR1 [56]. Thus, gene amplifications/rearrangements, rifampicin induction and probably other factors can cause MDR1 overexpression.

Defects in the apoptotic pathway might constitute an alternative mode of therapy resistance [57]. The p53 protein is an important regulator of the cell cycle and is sensitive to any DNA damage (genotoxic stress) caused during replication, in which case it will normally induce G1 arrest and/or apoptosis to prevent the production of defective cells. Drugs, which increase DNA damage, will therefore lead to p53-mediated cell death. Mutations in this gene are frequently observed in human cancers; loss of p53 function will allow cells with damaged DNA to continue replicating, which means resistance to DNA-damaging drugs. From a trial of ovarian carcinoma, Petty et al. [58] concluded that enhanced genomic instability due to p53 inactivation might increase the likelihood of development of resistance to CTX over time. In a variety of tumours, p53 deletion was reported to be associated with MDR [59]. Inactivated p53 almost completely suppressed the induction of apoptosis or severely delayed it. Johnson and Fan [60] found that reduced expression of p53 in human breast cancer cells modified response to paclitaxel and 5-FU. It has been suggested that reactivation of mutant p53 may be an effective way of inducing tumour cell death [61]. Susceptibility of tumour cells to programmed cell death is influenced by a series of proto-oncogenes and tumour suppressor genes. Other genes, such as h-ras and bcl-2/bax, involved in the apoptotic pathway, have also been implicated in resistance [62]. This type of resistance will affect a wider range (possibly all) of anticancer drugs. It also potentially increases the proportion of surviving mutant cells, which leads to greater tumour heterogeneity.

Enhanced DNA Repair

Another way that cells can become resistant, for example to cisplatin, is by developing an enhanced ability to remove cisplatin-DNA adducts and to repair cisplatin-induced lesions, through the action of DNA repair proteins. Levels of a nuclear protein called XPE-BF (xeroderma pigmentosum group E binding factor) were found to increase early in the development of cisplatin resistance. Excision repair cross-complementing protein (ERCC1) [63], most likely a DNA-binding protein, is another example of a DNA repair protein that may be involved in recognition of cisplatin damage; its expression is elevated in cisplatin-resistant cells compared with that in cells sensitive to cisplatin. The level of ERCC1 has been reported to increase as tumours become resistant to carboplatin [64]. Cross-resistance to carboplatin but not so much to either oxaliplatin or tetraplatin has been observed in cisplatin-resistant cells [65], so there are opportunities for second-line therapy. Resistance in non-seminomatous germ cell cancer cell lines was 1.7- to 2.2-fold with oxaliplatin compared to 3.9- to 6.1-fold with cisplatin [66].

Alterations in Target Molecules

It is possible that during the course of therapy, the target of that therapy could be modified in some way or even decrease to a level where it ceases to have any significant cellular influence and is therefore no longer a useful target to block. A fairly common example of this is encountered with anti-oestrogen (e.g. tamoxifen) therapy of breast cancer. Patients often undergo transition to an endocrine-resistant, from an initially responsive state, in which there is an apparent loss of oestrogen receptors in the resistant tumour cells. These patients no longer respond to tamoxifen treatment; however, their tumours can still, temporarily at least, be growth inhibited by inhibitors of oestrogen synthesis such as aromatase inhibitors [67, 68]. This is followed by a state of complete unresponsiveness to any endocrine manipulation. Presumably, the surviving cancer cells are no longer dependent on oestrogen for growth and continued attempts to block the now diminished original target, i.e. the oestrogen receptor, are fruitless.

Gene mutations are common in cancer cells; indeed it is this phenomenon, which gives that cell its remarkable properties. Cytotoxic drugs are aimed at disabling a component whose continued function is necessary for cell survival; cells that survive the treatment may do so by carrying a gene, for that target, which has mutated in such a way that it produces a protein that retains its activity but no longer binds to the drug, for stereochemical reasons, and is therefore not inhibited by it. The result is a functioning cell resistant to the drug. Resistance to the herbicide, glyphosate, can be induced in plants by transfection with a mutated aroA gene which encodes an enzyme necessary for synthesis of aromatic amino acids. The modified enzyme retains its essential catalytic activity but no longer binds glyphosate. Imatinib (Gleevac; STI571) is a tyrosine kinase inhibitor that induces apoptosis by preventing cell growth in cancerous cells by disabling the damaged bcr-abl receptors, preventing ATP binding. In clinical trials treating chronic myeloid leukaemia with STI571, Gorre et al. [69] found that patients in remission had reactivated bcr-abl activity; 3 of the 11 patients had amplified copies of the oncogenic bcr-abl gene. Two thirds of patients tested harboured a single-point mutation within the ATP-binding site of bcr-abl. Thus, the bcr-abl gene appears important in both the initiation and the maintenance of tumourigenicity. Identifying mutated alleles may help to detect drug-resistant clones prior to clinical relapse.

Topoisomerase, because of its vital role in DNA replication, is a favourite target for CTX. Mutations in this protein, which alter its nuclear localisation, render those cells insensitive to drugs such as etoposide designed to block the activity of topoisomerase II.

Target molecules may also disappear from cancer cells as a result of the loss of the corresponding gene. Chromosomal losses are a common feature of cancer. Duesberg et al. [70, 71] have suggested that it is the aneuploid nature of cancer cells that best explains the development of MDR. They conclude that the cause of this resistance is not simply the loss of drug-sensitive genes when a chromosome is lost, which can happen frequently given the large number of cells undergoing division in a tumour. Chromosome reassortment during mitosis is also capable of activating/inactivating numerous biochemical pathways that could disarm a variety of drugs. By contrast, the chances for a normal cell to lose or gain a chromosome during mitosis are undetectably low under most conditions and hence, it is not usual to see resistance to CTX in normal cells.

Penetration of drugs to their intended site of action is also a problem that is encountered and must be regarded as a form of resistance. The central portion of large tumours tends to have a poor blood supply and drugs will have limited access to this area. Therefore, CTX is more effective for smaller tumours and becomes less so as the tumour becomes larger. Treatment of brain tumours also faces the problems posed by the blood-brain barrier. This is a dynamic network of vessels, which restrict movement of molecules into the CNS. Many drugs are ineffective because they cannot pass through this barrier. Others, which may initially have been able to penetrate inside, may later be blocked due to the dynamic nature of this resistance. P-gp is an important component of this barrier and is present in high concentration on the apical surface of the endothelial cells [72–74]. While it restricts access of various potentially harmful pharmacologic agents to the brain, it also poses a problem in the treatment of brain diseases. Increasing the concentration of the drug to achieve entry into the brain to circumvent the action of P-gp presents problems of systemic toxicity. Begley [75] has reviewed the role of ABC transporters.

Genetic variations in transporter and metabolic enzymes are associated with differences in drug absorption, distribution, metabolism and excretion and are considered to be the major determinants of interindividual variability. Cytochrome p450s (cyp450) are a group of enzymes that alter the chemical structure of drugs, for elimination. Most drugs that are P-gp substrates are also Cyp3A4 substrates. These enzymes are located in the cells of the intestinal wall, endothelium, liver and other tissues. Genetic differences in cyp450 structure and expression lead to functional differences in drug and nutrient absorption and clearance. Over 50 cyp450 genes are expressed in different tissues of the human body, with more than a thousand identified in all species [76]. CYP3A4 is the most abundantly expressed P450 in human liver and small intestine [77] and is known to metabolise more than 120 different drugs. Hepatic expression of CYP3A4 is known to vary by more than 50-fold among individuals and in vivo CYP3A4 enzymatic function (drug clearance) varies by at least 20-fold. The genetic basis for this was reviewed by Lamba et al. [78]. The combination of P-gp and cyp450 largely determines drug absorption and biodistribution to tissues, limiting intestinal transcellular permeability, biliary disposition in hepatocytes, urinary elimination through the renal epithelium and placental transport. In 2003, Roche Diagnostics launched the Amplichip CYP450, the first pharmacogenomic micro-array for clinical applications. This tests for variants of cyp2D6, as this is missing in 7% of Caucasians and 2% of non-Caucasians. CYP2D6 is also hyperactive in 30% of East Africans. Therefore, CYP2D6 testing is considered more important than CYP3A as its expression varies more from one person to the next.

Metabolic Effects

Xenobiotics often modify high-density apolipoprotein. The result of this is to enhance hepatic drug elimination, reducing its effective plasma concentration. Another physiological response to the presence of drugs is the overexpression of drug-metabolising enzymes or carrier molecules; e.g. the increased production of glutathione [79] or ubiquitin contributes to inactivation of the drug by forming conjugates that are excreted. Resistance to cisplatin in ovarian carcinoma cells is associated with increased expression of dihydrodiol dehydrogenase [80]. In other cases, *underexpression* of drug-metabolising enzymes can reduce drug efficacy in situations where an administered inactive prodrug has to undergo catalytic conversion to an active form. Thus, arabinoside is required to be activated by the action of deoxycytidine kinase; loss or mutation of this enzyme will render this drug less effective [81].

Protein kinase C also plays an important role in both drug exclusion and apoptosis [82]. There are 12 or more different isoforms of this enzyme, which may regulate different cellular phosphorylation events. Drug-resistant breast cancer cells can have elevated activity, but the role of particular isoforms is undetermined.

Studies on breast cancer cells have also suggested involvement of the extracellular matrix in drug resistance [83]. It is known that apoptosis can be mediated by activation of the Fas/Fas-L pathway. These authors showed that ligation of b1 integrins by their extracellular matrix ligands inhibits apoptosis mediated by paclitaxel and vincristine, which activate this pathway. These agents act by causing release of mitochondrial cytochrome c. Integrin-mediated protection from the drug-induced apoptosis and inhibition of cytochrome c release were dependent on the activation of the PI 3 kinase/Akt pathway.

Growth Factors

Serum levels of interleukin (IL)-6 have been found to be elevated in patients with various types of cancer. Conze et al. [84] demonstrated that autocrine production of IL-6 by breast cancer cells could promote resistance to CTX. Cells sensitive to the CTX did not express appreciable IL-6, whereas high levels were detectable in multidrug-resistant cells. The mechanism of this resistance was attributed to the activation of the CCAAT enhancer-binding protein family of transcription factors and induction of MDR1 gene expression. They suggested that as some breast tumours contain IL-6-producing cells, this might have potential as a factor for predicting CTX resistance.

Song et al. [85] have shown that chemoresistance may be induced by extracellular factors in tumour-bearing organs. Comparing chemosensitivity and proteins in different tumours and different culture systems, they found elevated levels of acidic and basic fibroblast growth factors in the media of solid and metastatic tumours. These conditioned media induced broad-spectrum resistance to drugs (paclitaxel, doxorubicin and 5-FU) with diverse structures and mechanisms of action. Application in combination, of these two growth factors, could produce a 10-fold increase in drug resistance, whereas suramin, a known inhibitor of fibroblast growth factors, was able to reverse this resistance.

Drug Resistance Testing

Not all patients are necessarily resistant to CTX. Some tumours may be highly resistant to most cytotoxic drugs, while others may be quite sensitive. Therefore, it would be useful to have some tests, which could predict resistance and subsequent clinical failure. This would spare patients from the trauma of ineffective CTX. A company called IMPATH (Los Angeles, Calif., USA) has developed a cell culture drug resistance assay, which they claim can help in the selection of chemotherapeutic agents that have the greatest likelihood of being clinically effective. Their method essentially measures ^{3}H-thymidine uptake into cultured tumour cells, taken from fresh biopsy specimens, in the presence of various drugs; an algorithm applied to the data is used to determine the probability that a patient will respond to the various therapies tested in the assay. Kern and Weisenthal [86] reported that this assay used in clinical trials was highly accurate in demonstrating that cells extremely resistant in vitro were also resistant in vivo. The usefulness of such in vitro testing however remains a controversial matter and oncologists still prefer to rely on experience and the results of clinical trials with the drug. There has been much discussion on the subject [87].

Approaches to Overcoming Drug Resistance

As the mechanisms responsible for drug resistance become clearer, several strategies for countering this phenomenon are beginning to emerge. The most obvious response, and the one most widely employed, is to use combination drug therapy. The general rationale for choosing which drugs to combine is to use drugs which are active against the tumour when used individually; to combine drugs that have different modes and sites of action to produce complementary/synergistic rather than just additive effect; to combine drugs with minimally overlapping toxicities, allowing administration of maximally effective doses of each active agent to optimally schedule each drug, and to use drugs with narrowest possible cycle intervals necessary for bone marrow recovery. Drugs that are typically used in combination therapy for breast cancer are listed in table 2.

High-dose CTX can destroy the tumour cells before they have the opportunity to form resistant colonies, but this often comes at too great a cost to the patient. Less severe strategies are needed. As the action of the ABC transporters is thought the most significant reason for resistance, they have received the most attention. A large number of drugs have been identified that are pharmacological antagonists of particularly the P-gp. These include a variety of quite distinct groups of compounds, some of which are listed in table 3. Used in conjunction with the cytotoxic drugs already mentioned, these can potentiate their effects by allowing those drugs to remain in the target cells longer. These agents may be more properly regarded as modulators or reversing agents as even those that appear to be inhibitors of drug efflux may themselves be substrates, in so far as they can stimulate ATPase activity. The use of these in combination can subvert the cell's defence. For example, P-gp blocks absorption in the gut and may be considered part of the 'first pass effect'; it can act as a 'gatekeeper' for subsequent P450 cytochrome action. If one drug is a substrate of both P-gp and the cytochrome P450 subfamily IIIA (nifedipine oxidase) polypeptide 4 (CYP3A4; found in close proximity in the intestinal mucosa), and a second drug is added that is a modulator (e.g. erythromycin, mibefradil, ketoconazole), then the first drug will be allowed to increase its intracellular accumulation as a result of inhibition of the P-gp blockade. Since CYP3A4 is also inhibited, unmetabolised drug will enter the circulation. Another example is the entry of the antidiarrhoeal, loperamide, into the CNS (from which it is normally excluded) by the simultaneous administration of the P-gp modulator, quinidine [88]. The use of P-gp inhibitors may lead to undesirable consequences. Thus, the simultaneous use of cytotoxic drugs and agents that block P-gp function has raised questions of safety. In knockout mice with a genetic disruption of the mdr1a P-gp, much higher levels of substrate drugs accumulated in the brain, with markedly slower elimination from the circulation, resulting in dramatically increased toxicity to normal tissue [89].

Another mode of MDR reversal utilises monoclonal antibodies, several of which inhibit P-gp-mediated drug efflux in vitro [90, 91]. For in vivo applications, compa-

Table 2. Selected adjuvant CTX drug combinations used for breast cancer treatment

CMF	Cyclophosphamide, methotrexate, 5-FU
AC	Doxorubicin, cyclophosphamide
CAF	Cyclophosphamide, doxorubicin, 5-FU
AC→T	Doxorubicin, cyclophosphamide, paclitaxel
CEF	Cyclophosphamide, epirubicin, 5-FU
CMFVP	Cyclophosphamide, methotrexate, 5-FU, vincristine, prednisone
AT	Doxorubicin, docetaxel (Taxotere)
AC→Taxoter	Doxorubicin, cyclophosphamide, docetaxel

Table 3. Selected list of drugs that modulate P-gp

Immunosuppressant	Anti-arrhythmic agent
Cyclosporin A	Quinidine
Valspodar (PSC833)	Antifungal agent
HIV protease inhibitors	Ketoconazole
Ritonavir	Sedative
Saquinavir	Midazolam
Nelfinavir	Acridone carboxamide
Calcium channel blocker	GG918 (GF120918)
Verapamil	Peptide chemosensitiser
Progesterone antagonist	Reversin 121
Mifepristone (RU486)	Reversin 205
Anti-oestrogen	
Tamoxifen	

nies such as Immunomedics Inc. are developing polyspecific antibodies targeting both MDR protein and cancer epitopes. These antibodies may also be conjugated to cytotoxins, which can be internalised.

Various other means of circumventing resistance that are being explored are based on experimental observations described above: these include adjuvant treatment with suramin to reverse fibroblast growth factor-mediated resistance [85]; inhibition of p13 kinase/Akt to prevent resistance mediated through integrin involvement, and re-expression of wild-type tumour suppressor genes.

There have also been considerable technological advances that have allowed the identification of genes, which are potentially responsible for drug resistance, which had not been recognised previously using traditional approaches. Differential or subtractive analyses of gene expression in drug-sensitive and drug-resistant cell lines or tumours can now be performed with cDNA chip micro-arrays. This may facilitate identification of drug response genes whose expression is altered by the drug, and of signaling and metabolic pathways involved in drug activity.

More recent innovations to inhibit MDR proteins include transcriptional/translational inhibition through the introduction of antisense oligonucleotides or ribozymes [92, 93]. Interference [94] is another approach to knock out specific mRNAs and is rapidly gaining popularity; MDR genes would represent a good target. However, as many of these methods require gene targeting and transfer, they are unlikely to produce any really significant in vivo applications anytime soon, despite considerable progress in the general approach using animal models.

A more immediate strategy, termed 'metronomic dosing', is to simultaneously target the associated endothelial cells [95], which support tumour cell growth. This involves combining anti-angiogenic drugs with a more frequent regimen of cytotoxic agents; endothelial cells are more susceptible to CTX and a lower dose of CTX agents can be used, resulting in reduced side-effects [96]. Experiments in tumour-bearing mice have shown that administration of DC101, a blocker of vascular endothelial growth factor, and a submaximal tolerated dose of vinblastine induced effective tumour regression. Given individually, both agents were ineffective [97].

Targeting of growth factor receptors, which are overexpressed in breast cancer, with antibodies has been successfully employed for reducing tumour burden. The prime example of this is the use of trastuzumab aimed against the HER2 protein [98]. Cetuximab is another antibody that specifically blocks the epidermal growth factor receptor (EGFR), which is overexpressed in several cancers, particularly of the GI tract [99–101]. It has been successfully used in patients who were refractory to treatment with fluorouracil and irinotecan [102, 103]. The effectiveness of the combination of irinotecan and cetuximab in patients with irinotecan-refractory tumours [104] suggests that cetuximab may circumvent irinotecan resistance. These authors suggested that EGFR inhibition by cetuximab may overcome this resistance by abrogating drug efflux [105–109], restoring apoptosis [110] or impairing DNA repair activity [111, 112]. However, the efficacy of cetuximab may not be related just to the overall level of EGFR in the tumour, but perhaps to its phosphorylation state [113].

Anti-angiogenic agents such as bevacizumab, a humanized monoclonal antibody against VEGF, are also effective for metastatic colorectal cancer given in combination with irinotecan, fluorouracil and leucovorin [114]. It is suggested that in this case bevacizumab may improve delivery of CTX by altering tumour vasculature and decreasing the elevated interstitial pressure in tumours [115, 116].

Putting Resistance to Good Use

One of the major problems with CTX is its unwanted toxicity on bone marrow. Although this tissue recovers after cessation of treatment, the patient is severely immune compromised and has to be given additional protective treatment. The phenomenon of drug resistance could be turned to advantage for the patient's benefit. Stem cells taken from the patient could be transduced with a retroviral construct containing an MDR gene and used to repopulate the bone marrow, conferring resistance against subsequent exposure to CTX [93, 117]. Knockout experiments in mice have demonstrated targeted disruption of the MDR1b gene [118]. Several clinical trials are examining the possibility of delivering MDR1 gene-expressing vectors into cancer patients to induce chemoprotection or for in vivo selection of regenerating stem cells. Preliminary results seem to indicate that only modest levels of protection can be achieved in vivo, due to the very low percentages of patients' cells expressing MDR1 following hematopoietic reconstitution [119].

The old adage that prevention is better than cure applies as much to this situation. A concerted effort should be made to identify factors that can predict the onset of resistance before its clinical manifestation. This could allow the physician valuable lead time to alter either the combination of cytotoxic drugs or the form of the therapy. Attempts to devise methods of early detection of metastatic relapse have concentrated mainly on the development of very sensitive molecular assays for the detection of tumour-associated genes/proteins in circulating cells in the blood or micrometastases in the bone marrow [120–122]. The continual monitoring of cancer patients on therapy, for the traditional tumour markers such as carcino-embryonic antigen, has proved of limited use except in a few cancers; e.g. bHCG is a good indicator of early relapse in choriocarcinomas. The absence of a clearly defined molecule that is produced or is present only in cancer cells has greatly hindered these efforts. For the common adenocarcinomas, epithelial cell markers have been investigated on the premise that such cells are normally absent from blood, and if they appear in cancer patients, they must originate from a growing tumour. Without this early warning system, the problem of overcoming development of chemoresistance is quite considerable. In an ideal situation, therapy would be tailored to suit the individual at the outset; this is unlikely at least for the very near future, despite rapid progress in pharmacogenomics. In the meantime, a better understanding of the mechanisms of resistance will at least allow the physician to modulate the therapy on a need to do basis.

References

1 Roskoski R Jr: The ErbB/HER receptor protein-tyrosine kinases and cancer. Biochem Biophys Res Commun 2004;319:1–11.
2 Gross ME, Shazer RL, Agus DB: Targeting the HER-kinase axis in cancer. Semin Oncol 2004; 31:9–20.
3 Finn RS, Slamon DJ: Monoclonal antibody therapy for breast cancer: Herceptin. Cancer Chemother Biol Response Modif 2003;21: 223–233.
4 Herbst RS: Review of epidermal growth factor receptor biology. Int J Radiat Oncol Biol Phys 2004;59:21–26.
5 Gruner BA, Weitman SD: The folate receptor as a potential therapeutic anticancer target. Invest New Drugs 1999;16:205–219.
6 Reddy JA, Low PS: Folate-mediated targeting of therapeutic and imaging agents to cancers. Crit Rev Ther Drug Carrier Syst 1998;15:587–627.
7 Zamble DB, Lippard SJ: Cisplatin and DNA repair in cancer chemotherapy. Trends Biochem Sci 1995;20:435–439.
8 Hurley LH: DNA and its associated processes as targets for cancer therapy. Nat Rev Cancer 2002;2:188–200.
9 Kast C, Gros P: Topology mapping of the amino-terminal half of multidrug resistance-associated protein by epitope insertion and immunofluorescence. J Biol Chem 1997;272: 26479–26487.
10 Kast C, Gros P: Epitope insertion favors a six transmembrane domain model for the carboxy-terminal portion of the multidrug resistance-associated protein. Biochemistry 1998; 37:2305–2313.
11 Dean M, Rzhetsky A, Allikmets R: The human ATP-binding cassette (ABC) transporter superfamily. Genome Res 2001;11:1156–1166.
12 Juliano RL, Ling V: A surface glycoprotein modulating drug permeability in Chinese hamster ovary cell mutants. Biochim Biophys Acta 1976;455:152–162.
13 Chen CJ, Chin JE, Ueda K, Clark DP, Pastan I, Gottesman MM, Roninson IB: Internal duplication and homology with bacterial transport proteins in the mdr1 (P-glycoprotein) gene from multidrug-resistant human cells. Cell 1986;47:381–389.
14 Ueda K, Cardarelli C, Gottesman MM, Pastan I: Expression of a full-length cDNA for the human 'MDR1' gene confers resistance to colchicine, doxorubicin, and vinblastine. Proc Natl Acad Sci USA 1987;84:3004–3008.
15 Gottesman MM, Hrycyna CA, Schoenlein PV, Germann UA, Pastan I: Genetic analysis of the multidrug transporter. Annu Rev Genet 1995; 29:607–643.
16 Endicott JA, Ling V: The biochemistry of P-glycoprotein-mediated multidrug resistance. Ann Rev Biochem 1989;58:137–171.
17 Higgins CF: ABC transporters: From microorganisms to man. Annu Rev Cell Biol 1992;8: 67–113.
18 Gottesman MM, Pastan I, Ambudkar SV: P-glycoprotein and multidrug resistance. Curr Opin Genet Dev 1996;6:610–617.
19 Gottesman MM, Fojo T, Bates SE: Multidrug resistance in cancer: Role of ATP-dependent transporters. Nat Rev Cancer 2002;2:48–58.
20 Cole SP, Bhardwaj G, Gerlach JH, Mackie JE, Grant CE, Almquist KC, Stewart AJ, Kurz EU, Duncan AM, Deeley RG: Over-expression of a transporter gene in a multidrug-resistant human lung cancer cell line. Science 1992;258: 1650–1654.
21 Borst P, Evers R, Kool M, Wijnholds J: A family of drug transporters: The multidrug resistance-associated proteins. J Natl Cancer Inst 2000;92:1295–1302.
22 Cole SP, Deeley RG: Multidrug resistance mediated by the ATP-binding cassette transporter protein MRP. Bioessays 1998;20:931–940.
23 Litman T, Druley TE, Stein WD, Bates SE: From MDR to MXR: New understanding of multidrug resistance systems, their properties and clinical significance. Cell Mol Life Sci 2001;58:931–959.
24 Lecureur V, Sun D, Hargrove P, Schuetz EG, Kim RB, Lan LB, Schuetz JD: Cloning and expression of murine sister of P-glycoprotein reveals a more discriminating transporter than MDR1/P-glycoprotein. Mol Pharmacol 2000: 57:24–35.
25 Smith AJ, van Helvoort A, van Meer G, Szabo K, Welker E, Szakacs G, Varadi A, Sarkadi B, Borst P: MDR3 P-glycoprotein, a phosphatidylcholine translocase, transports several cytotoxic drugs and directly interacts with drugs as judged by interference with nucleotide trapping. J Biol Chem 2000;275:23530–23539.
26 Kool M, van der Inden M, de Haas M, Baas F, Borst P: Expression of human MRP6, a homologue of the multidrug resistance protein gene MRP1, in tissues and cancer cells. Cancer Res 1997;57:3537–3547.
27 Nooter K, Westerman AM, Flens MJ, Zaman GJ, Scheper RJ, van Wingerden KE, Burger H, Oostrum R, Boersma T, Sonneveld P: Expression of the multidrug resistance-associated protein (MRP) gene in human cancers. Clin Cancer Res 1995;111:1301–1310.
28 Homolya L, Varadi A, Sarkadi B: Multidrug resistance-associated proteins: Export pumps for conjugates with glutathione, glucoronate or sulfate. Biofactors 2003;17:103–114.
29 Haimeur A, Conseil G, Deeley RG, Cole SP: The MRP-related and BCRP/ABCG2 multidrug resistance proteins: Biology, substrate specificity and regulation. Curr Drug Metab 2004;5:21–53.
30 Zaman GJ, Lankelma J, van Tellingen O, Beijnen J, Dekker H, Paulusma C, Oude Elferink RP, Baas F, Borst P: Role of glutathione in the export of compounds from cells by the multidrug-resistance-associated protein. Proc Natl Acad Sci USA 1995;92:7690–7694.
31 Schinkel, AH: The physiological function of drug-transporting P-glycoproteins. Semin Cancer Biol 1997;8:161–170.

32 Thiebaut F, Tsuruo T, Hamada H, Gottesman MM, Pastan I, Willingham MC: Cellular localization of the multidrug-resistance gene product P-glycoprotein in normal human tissues. Proc Natl Acad Sci USA 1987;84:7735–7738.
33 Klimecki WT, Futscher BW, Grogan TM, Dalton WS: P-glycoprotein expression and function in circulating blood cells from normal volunteers. Blood 1994;83:2451–2458.
34 Gottesman MM, Pastan I, Ambudkar SV: P-glycoprotein and multidrug resistance. Curr Opin Genet Dev 1996;6:610–617.
35 Johnstone RW, Ruefli AA, Smyth MJ: Multiple physiological functions for multidrug transporter P-glycoprotein? Trends Biochem Sci 2000;25:1–6.
36 Johnstone RW, Ruefli AA, Tainton KM, Smyth MJ: A role for P-glycoprotein in regulating cell death. Leuk Lymphoma 2000;38:1–11.
37 Shapiro AB, Fox K, Lam P, Ling V: Stimulation of P-glycoprotein-mediated drug transport by prazosin and progesterone. Evidence for a third drug-binding site. Eur J Biochem 1999;259:841–850.
38 Sikic BI, Fisher GA, Lum BA, Halsey J, Beketic-Oreskovic L, Chen G: Modulation and prevention of multidrug resistance by inhibitors of P-glycoprotein. Cancer Chemother Pharmacol 1997;40(suppl):S13–S19.
39 Lin JH, Yamazaki M: Clinical relevance of P-glycoprotein in drug therapy. Drug Metab Rev 2003;35:417–454.
40 Penson RT, Oliva E, Skates SJ, Glyptis T, Fuller AF Jr, Goodman A, Seiden MV: Expression of multidrug resistance-1 protein inversely correlates with paclitaxel response and survival in ovarian cancer patients: A study in serial samples. Gynecol Oncol 2004;93:98–106.
41 van den Heuvel-Eibrink MM, Sonneveld P, Pieters R: The prognostic significance of membrane transport-associated multidrug resistance (MDR) proteins in leukemia. Int J Clin Pharmacol Ther 2000;38:94–110.
42 Wuchter C, Leonid K, Ruppert V, Schrappe M, Buchner T, Schoch C, Haferlach T, Harbott J, Ratei R, Dorken B, Ludwig WD: Clinical significance of P-glycoprotein expression and function for response to induction chemotherapy, relapse rate and overall survival in acute leukemia. Haematologica 2000;85:711–721.
43 Consoli U, Santonocito A, Stagno F, Fiumara P, Privitera A, Parisi G, Giustolisi GM, Pavone B, Palumbo GA, Di Raimondo F, Milone G, Guglielmo P, Giustolisi R: Multidrug resistance mechanisms in chronic lymphocytic leukaemia. Br J Haematol 2002;116:774–780.
44 Kumta SM, Zhu QS, Lee KM, Griffith J, Chow LT, Leung PC: Clinical significance of P-glycoprotein immunohistochemistry and doxorubicin binding assay in patients with osteosarcoma. Int Orthop 2001;25:279–282.
45 Leonessa F, Clarke R: ATP binding cassette transporters and drug resistance in breast cancer. Endocr Relat Cancer 2003;10:43–73.
46 Roepe PD: What is the precise role of human MDR 1 protein in chemotherapeutic drug resistance? Curr Pharm Des 2000;6:241–60.
47 Smyth MJ, Krasovskis E, Sutton VR, Johnstone RW: The drug efflux protein, P-glycoprotein, additionally protects drug-resistant tumor cells from multiple forms of caspase-dependent apoptosis. Proc Natl Acad Sci USA 1998; 12:7024–7029.
48 Henkart PA: ICE family proteases: Mediators of all apoptotic cell death? Immunity 1996;4: 195–201.
49 Johnstone RW, Cretney E, Smyth MJ: P-glycoprotein protects leukemia cells against caspase-dependent, but not caspase-independent, cell death. Blood 1999;93:1075–1085.
50 Ruefli AA, Smyth MJ, Johnstone RW: HMBA induces activation of a caspase-independent cell death pathway to overcome P-glycoprotein-mediated multidrug resistance. Blood 2000;95:2378–2385.
51 Ruefli AA, Tainton KM, Darcy PK, Smyth MJ, Johnstone RW: P-glycoprotein inhibits caspase-8 activation but not formation of the death inducing signal complex (disc) following Fas ligation. Cell Death Differ 2002;9:1266–1272.
52 Cullen KV, Davey RA, Davey MW: Drug resistance does not correlate with resistance to Fas-mediated apoptosis. Leuk Res 2001;25: 69–75.
53 Fromm MF, Kauffmann HM, Fritz P, Burk O, Kroemer HK, Warzok RW, Eichelbaum M, Siegmund W, Schrenk D: The effect of rifampin treatment on intestinal expression of human MRP transporters. Am J Pathol 2000; 157:1575–1580.
54 Greiner B, Eichelbaum M, Fritz P, Kreichgauer HP, von Richter O, Zundler J, Kroemer HK: The role of intestinal P-glycoprotein in the interaction of digoxin and rifampin. J Clin Invest 1999;104:147–153.
55 Mickley LA, Spengler BA, Knutsen TA, Biedler JL, Fojo T: Gene rearrangement: A novel mechanism for MDR-1 gene activation. J Clin Invest 1997;99:1947–1957.
56 Hoffmeyer S, Burk O, von Richter O, Arnold HP, Brockmoller J, Johne A, Cascorbi I, Gerloff T, Roots I, Eichelbaum M, Brinkmann U: Functional polymorphisms of the human multidrug-resistance gene: Multiple sequence variations and correlation of one allele with P-glycoprotein expression and activity in vivo. Proc Natl Acad Sci USA 2000;97:3473–3478.
57 Dive C, Hickman JA: Drug-target interactions: Only the first step in the commitment to a programmed cell death? Br J Cancer 1991;64: 192–196.
58 Petty R, Evans A, Duncan I, Kurbacher C and Cree I: Drug resistance in ovarian cancer – The role of p53. Pathol Oncol Res 1998;4:97–102.
59 Grunicke H: Oncogenes and drug resistance. Deutsche Krebsgesellschaft e.V. Materials, 23rd Congr, Berlin, June 8–12, 1998.
60 Johnson KR, Fan W: Reduced expression of p53 and p21WAF1/CIP1 sensitizes human breast cancer cells to paclitaxel and its combination with 5-fluorouracil. Anticancer Res 2002;22:3197–3204.
61 Selivanova G: Mutant p53: The loaded gun. Curr Opin Investig Drugs 2001;2:1136–1141.
62 Burger, H, Nooter K, Boersma AWM, Van Wingerden KE, Looijenga LHJ, Jochemsen AG, Stoter G: Distinct p53-independent apoptotic cell death signaling pathways in testicular germ cell tumour cell lines. Int J Cancer 1999; 81:629–628.
63 Rosell R, Taron M, Ariza A, Barnadas A, Mate JL, Reguart N, Margel M, Felip E, Mendez P, Garcia-Campelo R: Molecular predictors of response to chemotherapy in lung cancer. Semin Oncol 2004;31:20–27.
64 Chu G: Cellular responses to cisplatin. The roles of DNA-binding proteins and DNA repair. J Biol Chem 1994;269:787–790.
65 Rixe O, Ortuzar W, Alvarez M, Parker R, Reed E, Paull K, Fojo T: Oxaliplatin, tetraplatin, cisplatin, and carboplatin: Spectrum of activity indrug-resistant cell lines and in the cell lines of the National Cancer Institute's Anticancer Drug Screen panel. Biochem Pharmacol 1996; 52:1855–1865.
66 Dunn TA, Schmoll HJ, Grunwald V, Bokemeyer C, Casper J: Comparative cytotoxicity of oxaliplatin and cisplatin in non-seminomatous germ cell cancer cell lines. Invest New Drugs 1997;15:109–114.
67 Miller WR: Biological rationale for endocrine therapy in breast cancer. Best Pract Res Clin Endocrinol Metab 2004;18:1–32.
68 Campos SM: Aromatase inhibitors for breast cancer in postmenopausal women. Oncologist 2004;9:126–136.
69 Gorre ME, Mohammed M, Ellwood K, Hsu N, Paquette R, Rao PN, Sawyers CL: Clinical resistance to STI-571 cancer therapy caused by BCR-ABL gene mutation or amplification. Science 2001;293:876–880.
70 Duesberg P, Stindl R, Hehlmann R: Explaining the high mutation rates of cancer cells to drug and multidrug resistance by chromosome re-assortments that are catalyzed by aneuploidy. Proc Natl Acad Sci USA 2000;97:14295–14300.
71 Duesberg P, Stindl R, Hehlmann R: Origin of multidrug resistance in cells with and without multidrug resistance genes: Chromosome re-assortments catalyzed by aneuploidy. Proc Natl Acad Sci USA 2001;98:11283–11288.
72 Ramakrishnan P: The role of P-glycoprotein in the blood-brain barrier. Einstein Q J Biol Med 2003;19:160–165.
73 Tatsuta T, Naito M, Oh-Hara T, Sugawara I, Tsuruo T: Functional involvement of P-glycoprotein in blood-brain barrier. J Biol Chem 1992;267:20383–20391.
74 Hughes CS, Vaden SL, Manaugh CA, Price GS, Hudson LC: Modulation of doxorubicin concentration by cyclosporin A in brain and testicular barrier tissues expressing P-glycoprotein in rats. J Neurooncol 1998;37:45–54.
75 Begley DJ: ABC transporters and the blood-brain barrier. Curr Pharm Des 2004;10:1295–1312.
76 Cytochrome CYP450 standardised nomenclature committee website: http://drnelson.utmem.edu/CytochromeP450.html.

77 Waxman DJ: P450 gene induction by structurally diverse xenochemicals: Central role of nuclear receptors CAR, PXR, and PPAR. Arch Biochem Biophys 1999;369:11–23.
78 Lamba JK, Lin YS, Schuetz EG, Thummel KE: Genetic contribution to variable human CYP3A-mediated metabolism. Adv Drug Deliv Rev 2002;54:1271–1294.
79 Gamcsik MP, Dubay GR, Cox BR: Increased rate of glutathione synthesis from cystine in drug-resistant MCF-7 cells. Biochem Pharmacol 2002;63:843–851.
80 Deng HB, Parekh HK, Chow KC, Simpkins H: Increased expression of dihydrodiol dehydrogenase induces resistance to cisplatin in human ovarian carcinoma cells. J Biol Chem 2002; 277:15035–15043.
81 Dumontet C, Fabianowska-Majewska K, Mantincic D, Callet Bauchu E, Tigaud I, Gandhi V, Lepoivre M, Peters GJ, Rolland MO, Wyczechowska D, Fang X, Gazzo S, Voorn DA, Vanier-Viornery A, MacKey J: Common resistance mechanisms to deoxynucleoside analogues in variants of the human erythroleukaemic line K562. Br J Haematol 1999;106: 78–85.
82 Hofmann J: Protein kinase C isozymes as potential targets for anticancer therapy. Curr Cancer Drug Targets 2004;4:125–146.
83 Aoudjit F, Vuori K: Integrin signaling inhibits paclitaxel-induced apoptosis in breast cancer cells. Oncogene 2001;20:4995–5004.
84 Conze D, Weiss L, Regen PS, Bhushan A, Weaver D, Johnson P, Rincon M: Autocrine production of interleukin 6 causes multidrug resistance in breast cancer cells. Cancer Res 2001;61:8851–8858.
85 Song S, Wientjes MG, Gan Y, Au JL: Fibroblast growth factors: An epigenetic mechanism of broad spectrum resistance to anticancer drugs. Proc Natl Acad Sci USA 2000;97:8658–8663.
86 Kern DH, Weisenthal LM: Highly specific prediction of antineoplastic drug resistance with an in vitro assay using suprapharmacologic drug exposures. J Natl Cancer Inst 1990;82: 582–588.
87 McGuire WL, Kern DH, Von Hoff DD, Weisenthal LM: In vitro assays to predict drug sensitivity and drug resistance. A panel discussion. Breast Cancer Res Treat 1988;12:7–21.
88 Sadeque AJ, Wandel C, He H, Shah S, Wood AJ: Increased drug delivery to the brain by P-glycoprotein inhibition. Clin Pharmacol Ther 2000;68:231–237.
89 Schinkel AH, Mol CA, Wagenaar E, van Deemter L, Smit JJ, Borst P: Multidrug resistance and the role of P-glycoprotein knockout mice. Eur J Cancer 1995;31A:1295–1298.
90 Pearson JW, Fogler WE, Volker K, Usui N, Goldenberg SK, Gruys E, Riggs CW, Komschlies K, Wiltrout RH, Tsuruo T, et al: Reversal of drug resistance in a human colon cancer xenograft expressing MDR1 complementary DNA by in vivo administration of MRK-16 monoclonal antibody. J Natl Cancer Inst 1991; 83:1386–1391.
91 Mechetner EB, Roninson IB: Efficient inhibition of P-glycoprotein-mediated multidrug resistance with a monoclonal antibody. Proc Natl Acad Sci USA 1992;89:5824–5828.
92 Pawlak W, Zolnierek J, Sarosiek T, Szczylik C: Antisense therapy in cancer. Cancer Treat Rev 2000;26:333–350.
93 Gottesman MM, Fojo T, Bates SE: Multidrug resistance in cancer: Role of ATP-dependent transporters. Nat Rev Cancer 2002;2:48–58.
94 Elbashir SM, Harborth J, Lendeckel W, Yalcin A, Weber K, Tuschl T: Duplexes of 21-nucleotide RNAs mediate RNA interference in cultured mammalian cells. Nature 2000; 411:494–498.
95 Kerbel RS: Inhibition of tumor angiogenesis as a strategy to circumvent acquired resistance to anti-cancer therapeutic agents. Bioessays 1991;13:31–36.
96 Hanahan D, Bergers G, Bergsland E: Less is more, regularly: Metronomic dosing of cytotoxic drugs can target tumor angiogenesis in mice. J Clin Invest 2000;105:1045–1047.
97 Klement G, Baruchel S, Rak J, Man S, Clark K, Hicklin DJ, Bohlen P, Kerbel RS: Continuous low-dose therapy with vinblastine and VEGF receptor-2 antibody induces sustained tumor regression without overt toxicity. J Clin Invest 2000;105:R15–R24.
98 Slamon DJ, Leyland-Jones B, Shak S, Fuchs H, Paton V, Bajamonde A, Fleming T, Eiermann W, Wolter J, Pegram M, Baselga J, Norton L: Use of chemotherapy plus a monoclonal antibody against HER2 for metastatic breast cancer that overexpresses HER2. N Engl J Med 2001;344:783–792.
99 Messa C, Russo F, Caruso MG, Di Leo A: EGF, TGF-alpha, and EGF-R in human colorectal adenocarcinoma. Acta Oncol 1998; 37:285–289.
100 Porebska I, Harlozinska A, Bojarowski T: Expression of the tyrosine kinase activity growth factor receptors (EGFR, ERB B2, ERB B3) in colorectal adenocarcinomas and adenomas. Tumour Biol 2000;21:105–115.
101 Salomon DS, Brandt R, Ciardiello F, Normanno N: Epidermal growth factor-related peptides and their receptors in human malignancies. Crit Rev Oncol Hematol 1995;19: 183–232.
102 Saltz L, Rubin MS, Hochster HS: Cetuximab (IMC-C225) plus irinotecan (CPT-11) is active in CPT-11 refractory colorectal cancer that expresses epidermal growth factor receptor (abstract). Prog Proc Am Soc Clin Oncol 2001;20:3a.
103 Saltz LB, Meropol NJ, Loehrer PJ Sr, Needle MN, Kopit J, Mayer RJ: Phase II trial of cetuximab in patients with refractory colorectal cancer that expresses the epidermal growth factor receptor. J Clin Oncol 2004;22:1201–1208.
104 Cunningham D, Humblet Y, Siena S, Khayat D, Bleiberg H, Santoro A, Bets D, Mueser M, Harstrick A, Verslype C, Chau I, Van Cutsem E: Cetuximab monotherapy and cetuximab plus irinotecan in irinotecan-refractory metastatic colorectal cancer. N Engl J Med 2004; 351:337–345.
105 Chen ZS, Furukawa T, Sumizawa T: ATP-dependent efflux of CPT-11 and SN-38 by the multidrug resistance protein (MRP) and its inhibition by PAK-104P. Mol Pharmacol 1999;55:921–928.
106 Chu XY, Suzuki H, Ueda K, Kato Y, Akiyama S, Sugiyama Y: Active efflux of CPT-11 and its metabolites in human KB-derived cell lines. J Pharmacol Exp Ther 1999;288:735–741.
107 Meyers MB, Merluzzi VJ, Spengler BA, Biedler JL: Epidermal growth factor receptor is increased in multidrug-resistant Chinese hamster and mouse tumor cells. Proc Natl Acad Sci USA 1986;83:5521–5525.
108 Meyers MB, Yu P, Mendelsohn J: Crosstalk between epidermal growth factor receptor and P-glycoprotein in actinomycin D-resistant Chinese hamster lung cells. Biochem Pharmacol 1993;46:1841–1848.
109 Naruse I, Ohmori T, Ao Y, Fukumoto H, Kuroki T, Mori M, Saijo N, Nishio K: Antitumor activity of the selective epidermal growth factor receptor-tyrosine kinase inhibitor (EGFR-TKI) Iressa (ZD1839) in an EGFR-expressing multidrug-resistant cell line in vitro and in vivo. Int J Cancer 2002;98:310–315.
110 Sclabas GM, Fujioka S, Schmidt C, Fan Z, Evans DB, Chiao PJ: Restoring apoptosis in pancreatic cancer cells by targeting the nuclear factor-kappaB signaling pathway with the anti-epidermal growth factor antibody IMC-C225. J Gastrointest Surg 2003;7:37–43.
111 Bandyopadhyay D, Mandal M, Adam L, Mendelsohn J, Kumar R: Physical interaction between epidermal growth factor receptor and DNA-dependent protein kinase in mammalian cells. J Biol Chem 1998;273: 1568–1573.
112 Huang SM, Harari PM: Modulation of radiation response after epidermal growth factor receptor blockade in squamous cell carcinomas: Inhibition of damage repair, cell cycle kinetics, and tumor angiogenesis. Clin Cancer Res 2000;6:2166–2174.
113 Ciardiello F, Tortora G: Epidermal growth factor receptor (EGFR) as a target in cancer therapy: Understanding the role of receptor expression and other molecular determinants that could influence the response to anti-EGFR drugs. Eur J Cancer 2003;39:1348–1354.
114 Hurwitz H, Fehrenbacher L, Novotny W, Cartwright T, Hainsworth J, Heim W, Berlin J, Baron A, Griffing S, Holmgren E, Ferrara N, Fyfe G, Rogers B, Ross R, Kabbinavar F: Bevacizumab plus irinotecan, fluorouracil, and leucovorin for metastatic colorectal cancer. N Engl J Med 2004;350:2335–2342.

115 Jain RK: Normalizing tumor vasculature with anti-angiogenic therapy: A new paradigm for combination therapy. Nat Med 2001;7:987–989.
116 Willett CG, Boucher Y, di Tomaso E, Duda DG, Munn LL, Tong RT, Chung DC, Sahani DV, Kalva SP, Kozin SV, Mino M, Cohen KS, Scadden DT, Hartford AC, Fischman AJ, Clark JW, Ryan DP, Zhu AX, Blaszkowsky LS, Chen HX, Shellito PC, Lauwers GY, Jain RK: Direct evidence that the VEGF-specific antibody bevacizumab has antivascular effects in human rectal cancer. Nat Med 2004; 10:145–147.
117 Ambudkar SV, Dey S, Hrycyna CA, Ramachandra M, Pastan I, Gottesman MM: Biochemical, cellular, and pharmacological aspects of the multidrug transporter. Annu Rev Pharmacol Toxicol 1999;39:361–398.
118 Altuvia S, Stein WD, Goldenberg S, Kane SE, Pastan I, Gottesman MM: Targeted disruption of the mouse mdr1b gene reveals that steroid hormones enhance mdr gene expression. J Biol Chem 1993;268:27127–27132.
119 Sorrentino BP: Gene therapy to protect haematopoietic cells from cytotoxic cancer drugs. Nat Rev Cancer 2002;2:431–441.
120 Schoenfield A, Luqmani YA, Sinnett HD, Shousha S, Coombes RC: Keratin 19 mRNA measurement to detect micrometastases in lymph nodes in breast cancer patients. Br J Cancer 1996;74:1639–1642.
121 Pantel K, von Knebel Doeberitz M: Detection and clinical relevance of micrometastatic cancer cells. Curr Opin Oncol 2000;12: 95–101.
122 Zehentner BK: Detection of disseminated tumor cells: Strategies and diagnostic implications. Expert Rev Mol Diagn 2002;2: 41–48.

Review

Medical Principles and Practice

Med Princ Pract 2005;14(suppl 1):49–57
DOI: 10.1159/000086184

Received: June 2, 2004
Revised: December 21, 2004

Molecular Mechanisms of Organelle Biogenesis and Related Metabolic Diseases

Gursev S. Dhaunsi

Department of Pediatrics, Faculty of Medicine, Kuwait University, Kuwait

Key Words
Organelle biogenesis · Mitochondria · Lysosomes · Peroxisomes · Metabolic diseases

Abstract
Organelle biogenesis is regulated by transcriptional networks that modulate expression of specific genes encoding organellar proteins. Structural and functional specificity of organelles requires not only the transcription of specific genes and translation of resulting mRNAs, but also the transfer of encoded polypeptides to their site of function through signaling peptides. A defect in targeting of proteins to their subcellular site of function may not necessarily prevent biogenesis of the organelle, but would definitely lead to formation of a defective organelle with respect to that specific function. Several metabolic diseases are associated with dysfunction or defects in organelle biogenesis; among these, peroxisome biogenesis disorders, mitochondrial biogenesis defects and lysosomal storage disorders are an extensively studied group of genetic diseases where biogenesis of the organelle is compromised either due to a defect in assembly of the organelle itself or impaired import of matrix proteins into the organelle. Recent advances in biochemical and molecular aspects of biogenesis of subcellular organelles have not only unraveled the mechanisms for organization of cellular networks, but have also provided new insights into the development of metabolic diseases that are caused by disruption of organelle biogenesis. This article reviews the molecular mechanisms of biogenesis of mitochondria, lysosomes and peroxisomes in relation to the metabolic diseases of genetic or nongenetic origin.

Introduction

Complex organization of distinct membrane-bound compartments, called subcellular organelles, in eukaryotic cells forms the cellular basis of human physiology. Subcellular organelles interact with each other through their highly specialized metabolic activities and sustain various cellular functions. The amount, size and composition of each organelle in a cell are determined by the cellular metabolic needs according to its environment in the body. A complete set of organelles such as nucleus, mitochondria, endoplasmic reticulum, peroxisomes and lysosomes are always inherited maternally and proliferation, also termed organelle biogenesis, of these endowed organelles leads to an increased number and size of specialized membrane-bound cell compartments [1, 2].

Organelle biogenesis is regulated by transcriptional networks that modulate expression of specific genes encoding organellar proteins. Structural and functional specificity of organelles requires not only the transcription of specific genes and translation of resulting mRNAs, but also the transfer of encoded polypeptides to their site of function [3]. Proteins required for biogenesis of all sub-

KARGER
Fax +41 61 306 12 34
E-Mail karger@karger.ch
www.karger.com

1011–7571/05/0147–0049$22.00/0

Accessible online at:
www.karger.com/mpp

Dr. Gursev S. Dhaunsi
Department of Pediatrics, Faculty of Medicine, Kuwait University
PO Box 24923
13110 Safat (Kuwait)
Tel. +965 531 9486, Fax +965 533 8940, E-Mail dhaunsig@hsc.edu.kw

cellular organelles are encoded by nuclear genes except for mitochondria, where expression of both nuclear and mitochondrial genes is required for normal structure and metabolic functions of mitochondria [4, 5].

Several metabolic diseases are associated with dysfunction of organelle biogenesis and among these human diseases, there are three well-defined groups of deadly genetic disorders, namely mitochondrial disorders, lysosomal diseases and peroxisomal disorders, where functions of membrane-bound intracellular organelles are impaired [6–9]. Significant advances in biochemical and molecular aspects of biogenesis of subcellular organelles have been made which have unraveled the role of specific transcription factors in the expression of mitochondrial, lysosomal and peroxisomal proteins. Studies revealing the selective translocation of organellar proteins into the organelles through specific targeting signals have further clarified the complex process of organelle biogenesis [10, 11]. These recent developments in cell biology have not only unraveled the mechanisms for organization of cellular networks, but have also provided new insights into the development of metabolic diseases that are caused by disruption of organelle biogenesis.

Functions of various subcellular organelles are compromised not only in genetic disorders of organelle biogenesis, but several metabolic disturbances also occur during diseases of nongenetic origin such as infections, intoxications and drug treatments. Better understanding of biochemical and molecular aspects of organelle biogenesis strengthens the hope for correcting or preventing the impairment of metabolic activities during various pathological conditions of both genetic and nongenetic origin. This article aims to give an overview of our current knowledge about molecular mechanisms of biogenesis of mitochondria, lysosomes and peroxisomes, and inherited disorders in which these organelles are lacking or their functions are impaired due a defect in biogenesis.

Mitochondrial Biogenesis and Relevant Diseases

The mitochondria are the powerhouse of eukaryotic cells where aerobic ATP production during coupled oxidative phosphorylation is the key function along with numerous other biosynthetic and degradative reactions fundamental to cell function [12]. Recent advances in molecular cell biology have revealed that mitochondria also play a role in regulating the cellular life cycle through the release of cytochrome c, an important stimulator of apoptosis [13, 14]. The distinctive mitochondrial structure, which allows specialized functions involving different enzymes and reactions localized in discrete membranes and aqueous compartments, is the product of both synthesis of macromolecules within the mitochondria and the import of proteins and lipids synthesized outside the organelle. Mitochondria are maternally inherited and their growth is facilitated by synthesis and import of new mitochondrial components. Although the mitochondrion contains its own genetic material, the majority of its protein is derived from the nucleus, including the proteins involved in mitochondrial transcription and DNA replication. Distinct from the nuclear DNA, mammalian mitochondrial DNA (mtDNA) is a small circular molecule consisting of 16–18 kb that only encodes 13 mRNAs, 22 tRNAs and 2 rRNAs, of which all protein products function in the respiratory chain of the inner mitochondrial membrane. The nuclear genes for proteins to be targeted to the mitochondria are transcribed in the nucleus, translated in the cytosol, and translocated into the mitochondria, and this process of mitochondria biogenesis is regulated in a complex way through activation of specific transcription factors. Mitochondrial transcription factor A (Tfam) plays an important role in mtDNA transcription and replication [15, 16], and an increase in Tfam expression would precede alterations in mitochondrial gene expression. The promoter region of the human TFAM gene has a binding site for nuclear respiratory factor (NRF) that appears to be crucial in regulating the transcriptional activation of this gene [17]. The NRFs (NRF-1 and NRF-2) have been identified as nuclear transcription factors, and are common to a large number of nuclear-encoded mitochondrial proteins [18]. NRF-1 appears to bind to DNA as a homodimer where its C-terminal end activates transcription, while phosphorylation of NRF-1 on N-terminal serine residues enhances its DNA binding ability [19, 20]. In addition to its role in regulating Tfam expression, NRF-1 may furthermore influence mtDNA maintenance through its activation of the RNase mitochondrial RNA processing endonuclease. The RNase mitochondrial RNA processing is responsible for cleaving RNA transcripts to form the primers necessary for mtDNA replication [21]. Recently, a thermogenic coactivator, PGC-1, has also been shown to stimulate mitochondrial biogenesis by increasing the expression of NRF-1 and NRF-2. NRF-2 binding sites are present in the upstream regions of many nuclear-encoded mitochondrial proteins, including cytochrome c oxidase subunits IV and Vb, ATP synthase β-subunit, and human Tfam [17, 18]. Genes that encode mitochondrial oxidative phosphorylation proteins and Tfam also have binding sites for the transcription factor Sp1 in their promoter region [18, 22]. Sp1 is known to bind to GC-rich *cis*-elements [23, 24] and may compete with the zinc finger transcription factor Egr-1 for similar promoter sites [25, 26].

It is now well established that mitochondrial proteins are imported into the organelle posttranslationally. Mitochondria of the yeast *Saccharomyces cerevisiae* contain at least 750 different proteins, which perform diverse roles. Most of these proteins (approximately 99%) are translated on cytosolic ribosomes, and their import into mitochondria is essential for mitochondrial function. Further, the proteinaceous machineries of great complexity, the so-called translocases, in the mitochondrial membranes mediate the import of these proteins [27]. Following synthesis in the cytosol, mitochondrial preproteins are stabilized in an import-competent manner by cytosolic chaperones, such as the heat shock protein 70 (HSP70). This allows for the positively charged presequence to interact with acidic regions on the import receptor Tom20-Tom22 (translocase of the outer membrane). HSP70 is then released in an ATP-independent manner and the precursor is inserted into the outer mitochondrial membrane [28, 29]. In addition to the HSP70 chaperone pathway, preproteins containing internal targeting sequences interact with a chaperone mitochondrial import-stimulating factor that binds specifically to the import receptor Tom70-Tom37, and transfers the precursor to this subcomplex through the hydrolysis of ATP [28, 29]. The inner mitochondrial membrane uniquely contains the acidic phospholipid cardiolipin, which is believed to aid the import process [30]. The translocase protein complexes of the inner mitochondrial membrane are called the Tim proteins (translocase of the inner membrane) [31], which form a pore across the inner membrane consisting of Tim23 and Tim17, and guide the positively charged presequence across the polarized inner membrane [32, 33]. The mitochondrial inner membrane imports numerous proteins that span it multiple times using the membrane potential delta psi as the only external energy source. Rehling et al. [34] have shown that protein insertion complex (TIM22 complex), a twin-pore translocase, mediates the insertion of precursor proteins in a three-step process. After the precursor is tethered to the translocase without losing energy from the delta psi, two energy-requiring steps, namely action of delta psi on the precursor protein to promote its docking in the translocase complex and then, delta psi and an internal signal peptide, together induce rapid gating transitions in one pore and closing of the other pore and drive membrane insertion to completion. Upon arrival in the matrix, the presequence is cleaved by the mitochondrial processing peptidase, and the mature protein is folded into its proper conformation by the chaperones mtHSP70, HSP60 and/or chaperonin 10.

Activation of transcription factors and mitochondrial biogenesis are now known to be affected by a number of inherited genetic defects. A number of mitochondrial diseases associated with mtDNA defects ranging from point mutations to large-scale deletions or marked reduction in mtDNA quantity (depletion) form a group of neurogenetic disorders called neuromuscular diseases [35]. Point mutations in mtDNA that lead to mitochondrial disorders are maternally inherited and tend to cause a wide spectrum of clinical phenotypes [36]. Leigh's diseases and Leber's hereditary optical neuropathy are caused by point mutations in mtDNA genes and have a broad spectrum of clinical defects such as encephalomyopathies, ataxia, cardiomyopathy or sensorineural hearing loss [37, 38]. Major deletions in mtDNA have also been reported to occur that cause diseases of skeletal muscle, mitochondrial encephalomyopathies and cardiomyopathies. Kearns-Sayre syndrome, progressive external ophthalmoplegia and Pearson's marrow syndrome are the result of large-scale single mtDNA deletions and impairment of mitochondrial protein synthesis [39]. A significant reduction in mtDNA, also called depletion of mtDNA and characterized by a concomitant reduction in respiratory complex activities, has been reported to cause congenital or infantile myopathies. Studies investigating mitochondrial myopathies with depleted mtDNA have reported a corresponding reduction in Tfam protein and mRNA [40, 41]. On the other hand, mitochondrial diseases caused by an accumulation (duplication) of mtDNA are also known and muscle fibers from 2 patients with an accumulation of mtDNA have been shown to contain an elevated amount of Tfam protein [40] as well. The nuclear genome, which encodes most of the proteins involved in biogenesis of mitochondria, has also been linked to mtDNA defects and related myopathies [42]. Mutations in nuclear genes that encode structural subunits of respiratory oxidative phosphorylation complexes I–III have been reported to cause Leigh syndrome or Leigh syndrome-like disorders where patients are presented with lactic acidosis and progressive neurodegeneration of variable onset. Nuclear genes encoding proteins required in assembly of respiratory complexes or maintenance of mtDNA have also been linked to encephalocardiomyopathies.

As the new information on molecular mechanisms of mitochondrial biogenesis is rapidly emerging, the role of nuclear and mitochondrial genes in mitochondrial dysfunction and related neuromuscular disorders is becoming clearer. Recent reports [43, 44] on the regulation of mitochondrial biogenesis by cytokines such as nitric oxide propose a new insight into the complex process of mitochondrial biogenesis. The fact that NO, a versatile molecule in human pathophysiology due to its myriad of regulatory effects on different types of cells, promotes mitochondrial biogenesis leads us into a new direction for studying mechanisms of neuromuscular disorders and provides new hope for cure or prevention of diseases with mitochondrial dysfunction.

Lysosome Biogenesis and Metabolic Diseases

Lysosomes are membrane-bound cytoplasmic organelles that receive membrane traffic input from the biosynthetic, endocytic and autophagic pathways, and serve as major degradative compartments in the eukaryotic cells [45]. The degradative function of these organelles is carried out by more than 50 acid-dependent hydrolases (e.g. proteases, lipases, glycosidases) contained within its lumen [46]. The limiting membrane of the lysosome contains highly glycosylated proteins, lysosomal-associated membrane proteins (LAMPs) such as LAMP-1, LAMP-2, and CD63/LAMP-3, in addition to the proteins that mediate transport of ions, amino acids, and other solutes across the lysosomal membrane and contribute to the maintenance of an acidic luminal pH in the range of 4.6–5.0 [47]. The lack of cation-dependent and cation-independent mannose-6-phosphate receptors (MPRs) in lysosomal membrane distinguishes lysosomes from late endosomal compartments, and lysosomes share this characteristic with lysosome-related organelles, a group of cell type-specific nondegradative bodies that includes melanosomes, lytic granules, major histocompatibility complex class II compartments, platelet-dense granules, basophil granules, and neutrophil azurophil granules [48, 49]. The physiological importance of lysosomes and related organelles is evidenced by the myriad of diseases resulting from defects in their biogenesis and function [50, 51]. Lysosomal biogenesis is an orchestration of the structural and functional elements of the lysosome to form an integrated organelle and involves the synthesis, targeting, functional residence, and turnover of the proteins that comprise the lysosome. The question of how lysosomes form has captivated the interest of cell biologists for decades and the known molecular mechanisms of lysosome biogenesis are believed to involve vesicular delivery of newly synthesized lysosomal proteins from the trans-Golgi network (TGN), fusion of lysosomes with late endosomes and sorting of membrane proteins into lumenal vesicles [52]. A cascade of events that leads to biogenesis of lysosomes include the formation of early endosomes by coalescence of vesicles from the plasma membrane, the removal of recycling vesicles and addition of TGN-derived vesicles to convert these endosomes to late endosomes, and eventually to lysosomes [53, 54]. Alternatively, the early endosomes, late endosomes and lysosomes are considered stable compartments and transport proceeding from early endosomes through an endosomal carrier vesicle with characteristics of a multivesicular body to late endosomes leads to mature lysosomes [55, 56].

Though different mechanisms have been proposed, yet the complexity of events involved in lysosomal biogenesis remains unresolved. For any model of lysosomal biogenesis to work, a mechanism for selective sorting of materials bound for lysosomes must exist and most newly synthesized lysosomal proteins are directed to lysosomes through an MPR-dependent or MPR-independent fashion [57]. The sorting of MPRs and lysosomal membrane proteins is mediated by signals present in the cytoplasmic domains that are recognized by cytosolic molecules recruited onto membranes at sites of vesicle formation (e.g. the TGN, plasma membrane and endosomes). A number of heterotetrameric adaptor protein (AP) complexes, AP-1, AP-2, AP-3 and AP-4, mediate transport of MPRs and lysosomal integral membrane proteins from the TGN to endosomes [58, 59]. AP-1 has been suggested to play a role in recycling MPRs from endosomes to the TGN. Two unrelated connector molecules, TIP47 [60] and PACS-1 [61], have also been implicated in recycling of MPRs to the TGN. Recent studies have identified a novel family of coat proteins, termed GGAs, which may be involved in biosynthetic protein transport from the TGN to the endosomal/lysosomal system. The recruitment of GGAs to the TGN appears to be mediated by members of the ADP-ribosylation factor family of proteins. Once carrier vesicles are formed and cargo proteins sorted into them, the vesicles need to be targeted and eventually fuse with acceptor compartments. GTP-binding proteins, called rabs, found at the plasma membrane and on organelles of secretory nature appear to be required for transport of MPRs to the lysosome [62]. Targeting and membrane fusion also require vesicle-associated membrane proteins collectively known as N-ethylmaleimide-sensitive factor attachment protein receptors, or SNAREs, and are found on both vesicles (v-SNAREs) and target membranes (t-SNAREs) [63]. The degree to which these, or other, mammalian SNAREs are required for sorting to the mature lysosome itself remains uncertain. Recent studies on lysosomal targeting have revealed a variety of proteins as components of the lysosome biogenesis machinery, and the establishment of physical and/or functional relationships between them will undoubtedly clarify understanding of lysosome biogenesis.

Advances in understanding the complexities of lysosomal biogenesis, lysosomal hydrolase sorting and hydrolytic requirements have led to the possibility of a successful enzyme therapy for lysosomal storage disorders, a group of over 45 distinct genetic diseases resulting from a deficiency of a particular lysosomal protein or, in a few cases, from nonlysosomal proteins that are involved in lysosomal biogenesis [64, 65]. Lysosomal enzymes sialidase (α-neuraminidase), β-galactosidase, and N-acetylaminogalacto-6-sulfate sulfatase are involved in the catabolism of glycolipids, glycoproteins, and oligosaccharides. Their functional activity in the cell depends on their association in a multienzyme complex with lysosomal carboxypeptidase, and cathepsin A and the integrity of the complex plays a crucial role at different stages of biogenesis of lysosomal enzymes, including intracellular sorting and proteolytic processing of their precursors [66]. The complex plays a protective role for all components, extending their half-life in the lysosome from several hours to several days, and for sialidase, the association with cathepsin A is also necessary for the expression of enzymatic activity. The disintegration of the complex due to genetic mutations in its components results in their functional deficiency and causes severe metabolic disorders: sialidosis (mutations in sialidase), GM1 gangliosidosis and Morquio disease type B (mutations in β-galactosidase), galactosialidosis (mutations in cathepsin A), and Morquio disease type A (mutations in N-acetylaminogalacto-6-sulfate sulfatase) [67–70].

Most of the properties of lysosomes are shared with a group of cell type-specific compartments referred to as 'lysosome-related organelles', which include melanosomes, lytic granules, major histocompatibility complex class II compartments, platelet dense granules, basophil granules and azurophil granules. In addition to lysosomal proteins, these organelles contain cell type-specific components that are responsible for their specialized functions. Abnormalities in both lysosomes and lysosome-related organelles have been observed in human genetic diseases such as the Chediak-Higashi and Hermansky-Pudlak syndromes [71–72]. Identification of genes mutated in these human diseases is beginning to shed light on the molecular machinery involved in the biogenesis of lysosomes and lysosome-related organelles.

Peroxisomes and Peroxisomal Biogenesis Disorders

Peroxisomes, the single-membrane-bound cell organelles also called microbodies, are now known to contain more than 60 different enzymes that carry out a number of important metabolic pathways including β-oxidation of very long chain fatty acids, synthesis of plasmalogens and degradation of reactive oxygen species [73, 74]. Metabolic functions of peroxisomes are vital for cell survival and the importance of peroxisomes in human health was highlighted in the late 1980s with recognition of a number of inherited fatal metabolic diseases with dysfunction of this organelle [75, 76]. Peroxisomal disorders, a relatively new category of inherited metabolic diseases, occur due to a defect in either the assembly of the organelle or import of peroxisomal proteins into the organelle. Biogenesis of peroxisomes maintains the required number and size of peroxisomes in the tissues according to the metabolic milieu [77]. Peroxisome biogenesis conceptually consists of (a) the formation of the peroxisomal membrane, (b) the import of proteins into the peroxisomal matrix and (c) proliferation of the organelles [78]. As a result of combined genetic and biochemical approaches, 32 genes have been identified that encode proteins involved in peroxisome biogenesis [79, 80]. These genes are called *PEX* genes and their protein products are termed peroxins. Elaborate peroxin complexes form the receptor machinery, which in a concerted action of the components transports folded, even oligomeric matrix proteins across the peroxisomal membrane. The past decade has significantly improved our knowledge of peroxins, which can now be clearly divided into two distinct groups: those involved exclusively in peroxisome matrix protein import, and those involved in peroxisome membrane biogenesis.

Of the above steps of peroxisome biogenesis, the most is understood about matrix protein import. Matrix proteins are encoded by nuclear genes, synthesized on free ribosomes in the cytoplasm and imported into the peroxisome posttranslationally. Two targeting signals for peroxisomal matrix proteins have been identified [81]. The most commonly used signal is the peroxisomal targeting

Table 1. Summary of classical mitochondrial, lysosomal and peroxisomal biogenesis disorders, their clinical characteristics and related defective genes

Inherited diseases	Clinical symptoms	Gene(s)
Mitochondrial biogenesis disorders		
Leigh's syndrome	bilateral lesions in basal ganglia, ataxia, seizures, hypotonia, lactic acidosis, developmental delay, hypertrophic cardiomyopathy	*mtDNA* deletion
Progressive external ophthalmoplegia	hypotonia, developmental delay, lactic acidosis, progressive ophthalmoplegia (external)	*mtDNA* deletion
Lysosomal biogenesis disorders		
Chediak-Higashi syndrome	progressive neurological impairment, bleeding diathesis, severe immunological deficiency, hypopigmentation of skin, hair and eyes	*CHS1*
Hermansky-Pudlak syndrome	oculocutaneous albinism, ceroid lipofuscinosis, prolonged bleeding	*HPS1*
Peroxisomal biogenesis disorders		
Zellweger syndrome	dysmorphic features, hypotonia, developmental delay, seizures, failure to thrive, hepatomegaly, psychomotor retardation	*PEX1*

signal 1, or PTS1, which consists of a conserved tripeptide at the extreme C terminus of a protein of consensus sequence Ser-Lys-Leu-COOH. The second targeting signal is the peroxisomal targeting signal 2, or PTS2, which must be present at the N terminus of a protein, and thiolase presents the best example of a peroxisomal protein with PTS2. The targeting signals are recognized by specific receptor proteins, the peroxins (PEX), present on the peroxisomal membrane or in the cytosol as carriers. PEX5 recognizes PTS1, while PEX7 binds PTS2, and these receptors bind their cargo in the cytoplasm, possibly with the aid of chaperone proteins [82]. Very little is known about how the receptors transport their cargo to the peroxisome. The receptors have been reported to interact with PEX13, suggesting that PEX13 is a member of the receptor docking complex. PEX14 has also been reported as equally important for the docking process of PTS receptors. PEX17 is also thought to be part of the docking complex and associates directly with PEX14. PEX10 and PEX12 are integral peroxisome membrane proteins, which have been shown to directly interact and are members of the putative translocation apparatus, along with PEX2 and PEX8. PEX3 has been shown to act as a putative peroxisomal receptor for PEX19. PEX12 and PEX8 have both been demonstrated to interact with PEX5. PEX5 has been demonstrated to be a cycling receptor and this recycling is thought to be mediated by the AAA ATPases PEX1 and PEX6, the ubiquitin-conjugating enzyme PEX4, and the putative docking site for PEX4, PEX22, PEX15 and PEX27 have recently been suggested to also act as binding partners for AAA peroxins at the peroxisomal membrane [83].

To date, only 3 peroxins have been directly implicated in the formation of peroxisomal membranes during peroxisome biogenesis. PEX3, PEX16, and PEX19 are the 3 proteins identified to be involved in the process of the formation of peroxisome membranes de novo [84]. PEX19 is the only peroxin known to directly bind the majority of known peroxisomal membrane proteins, and there are possibly other cytoplasmic proteins which aid PEX19 in the cytoplasmic portion of the peroxisome biogenesis process (i.e. chaperone proteins), but these proteins have not yet been identified. PEX11 is considered to be required for peroxisome proliferation since overexpression of PEX11 causes massive proliferation of peroxisomes. PEX11b is expressed constitutively, suggesting a general role in the maintenance of peroxisome number, whereas PEX11a is inducible, and can cause peroxisome proliferation in response to external stimuli [85].

Disorders of peroxisome biogenesis are multiorgan diseases and the nervous system is implicated in most of them. On a genetic basis, complementation of fibroblasts has classified peroxisome biogenesis disorders [86] into 12 complementation groups. Identification of peroxisome biogenesis genes (PEX) has, however, led to a new association of peroxisome biogenesis disorders with gene defects. Zellweger syndrome (also referred to as cerebrohepatorenal syndrome), neonatal adrenoleukodystrophy, and infantile Refsum disease along with rhizomelic chondrodysplasia punctata are autosomal recessive diseases collectively called the peroxisome biogenesis disorders. The biochemical phenotype of these disorders is reflective of the defect in peroxisome biogenesis as demonstrated by increased levels of very long chain fatty acid and phytanic acid, and reduced plasmalogen levels [87]. There are also high levels of the bile acid intermediates as well as increased levels of the lysine metabolite pipecolic acid. Zellweger syndrome belongs to complementation group 1, which is caused by mutations in PEX1 [88]. Within this complementation group, there are 2 common mutations, which together comprise 60% of the alleles in the complementation group. One is a mutation which causes the amino acid change G843D, and this is associated with the milder phenotypes of the Zellweger spectrum. The other, a frameshift in exon 13 of *PEX1,* is associated with the severe phenotype of Zellweger syndrome [89]. Rhizomelic chondrodysplasia punctata results from a defect in PEX7, the PTS2 receptor [90]. Because this receptor is only required for the import of PTS2-containing proteins, of which there are only 3 known in humans (thiolase, phytanoyl-CoA hydroxylase, and dihydroxyacetonephosphate synthase), a defect in this protein produces a very different clinical phenotype from that of the Zellweger spectrum. The biochemical phenotype of rhizomelic chondrodysplasia punctata also reflects the molecular basis of the disease [91]. Patients have a defect in branched-chain fatty acid metabolism as well as a defect in plasmalogen biosynthesis, but lack the other biochemical defects present in the Zellweger spectrum.

Primary hyperoxaluria type I [92] presents a unique peroxisomal disorder where targeting of alanine:glyoxylate aminotransferase, a peroxisomal enzyme, to mitochondria instead of peroxisome leads to urolithiasis and/or nephrocalcinosis.

In summary, molecular organization of subcellular organelles is key to their specific metabolic functions in the cell. Biogenesis of subcellular organelles determines the cellular metabolic state, which in turn maintains the physiologic functions of the tissue, leading to a pathological condition. Biogenesis of mitochondria, peroxisomes, lysosomes, endoplasmic reticulum and other subcellular organelles is tightly controlled by a complex network of events that involve the transcriptional regulation of the expression of organellar proteins, their posttranscriptional modifications, import into the organelle for structural organization and their intermolecular interaction. A defect in any of the components of this network leads to a serious pathological state of mitochondrial or lysosomal or peroxisomal diseases. Recent information on molecular mechanisms of organelle biogenesis has unraveled to some extent the complexity of interorganellar interactions as well, where an abnormal biogenesis of one organelle will affect biogenesis of the other organelle(s). A better understanding of the molecular mechanisms of organellar biogenesis has not only unraveled the pathogenic mechanisms of metabolic disorders but has also provided new tools for correcting the metabolic defects and related complications.

References

1 Nunnari J, Walter P: Regulation of organelle biogenesis. Cell 1996;84:389–394.
2 Warren G, Wickner W: Organelle inheritance. Cell 1996;84:395–400.
3 Hill C, Treisman R: Transcriptional regulation by extracellular signals: Mechanisms and specificity. Cell 1995;80:199–211.
4 Palade G: Membrane biogenesis: An overview. Methods Enzymol 1983;96:29–55.
5 Anderson S, Bankier AT, Barrel BG, de Bruijn MH, Coulson AR, Drouin J, Eperon IC, et al: Sequence and organization of the human mitochondrial genome. Nature 1981;290:457–465.
6 Marin-Garcia J, Goldenthal MJ: Mitochondrial biogenesis defects and neuromuscular disorders. Pediatr Neurol 2000;22:122–129.
7 Lazarow PB, Fujiki Y: Biogenesis of peroxisomes. Annu Rev Cell Biol 1986;1:489–530.
8 Lazarow PB, Moser HW: Disorders of peroxisome biogenesis; in Scriver CR, et al (eds): The Metabolic and Molecular Bases of Inherited Disease. New York, McGraw-Hill, 1995, pp 2287–2324.
9 Tager JM: Inborn errors of cellular organelles: An overview. J Inherit Metab Dis 1987;10 (Suppl1):3–10.
10 Walter P, Johnson AE: Signal sequence recognition and protein targeting to the endoplasmic reticulum membrane. Annu Rev Cell Biol 1994;10:87–119.
11 Ryan KR, Jensen RE: Protein translocation across mitochondrial membranes: What a long, strange trip it is. Cell 1995;83:517–519.
12 Senior AE: ATP synthesis by oxidative phosphorylation. Physiol Rev 1988;68:177–230.
13 Orrenius S: Mitochondrial regulation of apoptotic cell death. Toxicol Lett 2004;149:19–23.
14 Kluck RM, Bossy-Wetzel E, Green DR, Newmeyer DD: The release of cytochrome c from mitochondria: A primary site for Bcl-2 regulation of apoptosis. Science 1997;275:1132–1136.
15 Chang DD, Clayton DA: Priming of human mitochondrial DNA replication occurs at the light-strand promoter. Proc Natl Acad Sci USA 1985;82:351–355.
16 Fisher RP, Clayton DA: Purification and characterization of human mitochondrial transcription factor 1. Mol Cell Biol 1988;8:3496–3509.
17 Virbasius JV, Scarpulla RC: Activation of the human mitochondrial transcription factor A gene by nuclear respiratory factors: A potential regulatory link between nuclear and mitochondrial gene expression in organelle biogenesis. Proc Natl Acad Sci USA 1994;91:1309–1313.

18 Scarpulla R: Nuclear control of respiratory chain expression in mammalian cells. J Bioenerg Biomembr 1997;29:109–119.
19 Gugneja S, Scarpulla RC: Serine phosphorylation within a concise amino-terminal domain in nuclear respiratory factor 1 enhances DNA binding. J Biol Chem 1997;272:18732–18739.
20 Gugnega S, Vibasius CA, Scarpulla RC: Nuclear respiratory factors 1 and 2 utilize similar glutamine-containing clusters of hydrophobic residues to activate transcription. Mol Cell Biol 1996;16:5708–5716.
21 Clayton DA: Transcription and replication of animal mitochondrial DNAs. Int Rev Cytol 1992;141:217–232.
22 Lenka N, Vijayasarathy C, Mullick J, Avadhani NG: Structural organization and transcription regulation of nuclear genes encoding the mammalian cytochrome c oxidase complex. Prog Nucleic Acid Res Mol Biol 1998;61:309–344.
23 Berg JM: Sp1 and the subfamily of zinc finger proteins with guanine-rich binding sites. Proc Natl Acad Sci USA 1992;89:11109–11110.
24 Phillipsen S, Suske G: A tale of three fingers: The family of mammalian Sp/XKLF transcription factors. Nucleic Acids Res 1999;27:2991–3000.
25 Huang RP, Fan Y, Ni Z, Mercola D, Adamson ED: Reciprocal modulation between Sp1 and Egr-1. J Cell Biochem 1997;66:489–499.
26 Silverman ES, Khachigian LM, Lindner V, Williams AJ, Collins T: Inducible PDGF A-chain transcription in smooth muscle cells is mediated by Egr-1 displacement of Sp1 and Sp3. Am J Physiol 1997;273:H1415–H1426.
27 Chacinska A, Rehling P: Moving proteins from the cytosol into mitochondria. Biochem Soc Trans 2004;32:774–776.
28 Komiya T, Sakaguchi M, Mihara K: Cytoplasmic chaperones determine the targeting pathway of precursor proteins to mitochondria. EMBO J 1996;15:399–407.
29 Komiya T, Rospert S, Schatz G, Mihara K: Binding of mitochondrial precursor proteins to the cytoplasmic domains of the import receptors Tom70 and Tom20 is determined by cytoplasmic chaperones. EMBO J 1997;16:4267–4275.
30 Eilers M, Endo T, Schatz G: Adriamycin, a drug interacting with acidic phospholipids, blocks import of precursor proteins by isolated yeast mitochondria. J Biol Chem 1989;264:2945–2950.
31 Pfanner N, Douglas MG, Endo T, Hoogenraad NJ, Jensen RE, Meijer M, Neupert W, Schatz G, Schmitz UK, Shore GC: Uniform nomenclature for the protein transport machinery of the mitochondrial membranes. Trends Biochem Sci 1996;21:51–52.
32 Pfanner N, Meijer M: Mitochondrial biogenesis: The Tom and Tim machine. Curr Biol 1997;7:R100–R103.
33 Schatz G: The protein import system of mitochondria. J Biol Chem 1996:271:31763–31766.
34 Rehling P, Brandner K, Pfanner N: Mitochondrial import and the twin-pore translocase. Nat Rev Mol Cell Biol 2004;5:519–530.
35 Larsson N, Clayton DA: Molecular genetic aspects of human mitochondrial disorders. Annu Rev Genet 1995;29:151–178.
36 Schon EA, Bonilla E, DiMauro S: Mitochondrial DNA mutations and pathogenesis. J Bioenerg Biomembr 1997;29:131–149.
37 Corona P, Lamantea E, Greco M: Novel heteroplasmic mtDNA mutation in a family with heterogenous clinical presentations. Ann Neurol 2002;51:118–122.
38 Santorelli FM, Mak SC, El-Schahawi M: Maternally inherited cardiomyopathy and hearing loss associated with a novel mutation in the mitochondrial tRNA(Lys) gene (G8363A). Am J Hum Genet 1996;58.933–939.
39 Holt IJ, Harding AE, Morgan-Hughes JA: Deletions of mtDNA in patients with mitochondrial myopathies. Nature 1988;331:717–719.
40 Larsson N, Oldfors A, Holme E, Clayton DA: Low levels of mitochondrial transcription factor A in mitochondrial DNA depletion. Biochem Biophys Res Commun 1994;200:1374–1381.
41 Poulton J, Morten K, Freeman-Emmerson C, Potter C, Sewry C, Dubowitz V, Kidd H, Stephenson J, Whitehouse W, Hansen FJ, Parisi M, Brown G: Deficiency of the human mitochondrial transcription factor h-mtTFA in infantile mitochondrial myopathy is associated with mtDNA depletion. Hum Mol Genet 1994;3:1763–1769.
42 Goffart S, Wiesner RJ: Regulation and co-ordination of nuclear gene expression during mitochondrial biogenesis. Exp Physiol 2003;88:33–40.
43 Nisoli E, Clementi E, Paolucci C, Cozzi V, Tonello C, Sciorati C, Bracale R, Valero A, Francolini M, Moncada S, Carruba MO: Mitochondrial biogenesis in mammals: The role of endogenous nitric oxide. Science 2003;299:896–899.
44 Leary SC, Shoubridge EA: Mitochondrial biogenesis: Which part of 'NO' do we understand? Bioessays 2003;25:538–541.
45 Griffiths G: What's special about secretory lysosomes? Semin Cell Dev Biol 2002;13:279–84.
46 de Duve C: The lysosome concept; in de Reuck AVS, Cameron MP (eds): Lysosomes. Boston, Little Brown, 1963, pp 1–35.
47 Eskelinen EL, Tanaka Y, Saftig P: At the acidic edge: Emerging functions for lysosomal membrane proteins. Trends Cell Biol 2003;13:137–145.
48 Orlow SJ: Melanosomes are specialized members of the lysosomal lineage of organelles. J Invest Dermatol 1995;105:3–7.
49 Griffiths GM, Argon Y: Structure and biogenesis of lytic granules. Curr Top Microbiol Immunol 1995;198:39–58.
50 Spritz RA: Multi-organellar disorders of pigmentation: Tied up in traffic. Clin Genet 1999;55:309–317.
51 Meikle PJ, Hopwood JJ: Lysosomal storage disorders: Emerging therapeutic options require early diagnosis. Eur J Pediatr 2003;162(Suppl 1):S34–S37.
52 Mullins C, Bonifacino JS: The molecular machinery for lysosome biogenesis. Bioessays 2001;23:333–343.
53 Murphy RF: Maturation models for endosome and lysosome biogenesis. Trends Cell Biol 1991;1:77–82.
54 Storrie B, Desjardins M: The biogenesis of lysosomes: Is it a kiss and run, continous fusion and fission process? Bioessays 1996;18:895–903.
55 Dell'Angelica EC, Mullins C, Caplan S, Bonifacino JS: Lysosome-related organelles. FASEB J 2000;14:1265–1278.
56 Luzio JP, Rous BA, Bright NA, Pryor PR, Mullock BM, Piper RC: Lysosome-endosome fusion and lysosome biogenesis. J Cell Sci 2000;113:1515–1524.
57 Hunzikar W, Geuze HJ: Intracellular trafficking of lysosomal membrane proteins. Bioessays 1996;18:379–389.
58 Bonifacino JS, Dell'Angelica EC: Molecular bases for the recognition of tyrosine based sorting signals. J Cell Biol 1999;145:923–926.
59 Kirchhausen T: Adaptors for clathrin-mediated traffic. Annu Rev Cell Dev Biol 1999;15:705–732.
60 Diaz E, Pfeffer SR: YIP47: A cargo selection device for mannose-6-phosphate receptor trafficking. Cell 1998;93:433–443.
61 Wan L, Molloy SS, Thomas L, Liu G, Xiang Y, Rybak SL, Thomas G: PACS-1 defines a novel gene family of cytosolic sorting proteins required for trans-Golgi network localization. Cell 1998;94:205–216.
62 Press B, Feng Y, Hoflack B, Wandinger-Ness A: Mutant Rab7 causes the accumulation of cathepsin D and cation-independent mannose-6-phosphate receptor in an early endocytic compartment. J Cell Biol 1998;140:1075–1089.
63 Pfeffer SR: Transport-vesicle targeting: Tethers before SNAREs. Nat Cell Biol 1999;1:E17–E22.
64 Grabowski GA, Hopkin RJ: Enzyme therapy for lysosomal storage disease: Principles, practice, and prospects. Annu Rev Genomics Hum Genet 2003;4:403–436.
65 Meikle PJ, Fuller M, Hopwood JJ: Mass spectrometry in the study of lysosomal storage disorders. Cell Mol Biol 2003;49:769–777.
66 Pshezhetsky AV, Ashmarina M: Lysosomal multienzyme complex: Biochemistry, genetics, and molecular pathophysiology. Prog Nucleic Acid Res Mol Biol 2001;69:81–114.
67 Strehle EM: Sialic acid storage disease and related disorders. Genet Test 2003;7:113–121.
68 Callahan JW: Molecular basis of GM1 gangliosidosis and Morquio disease, type B. Structure-function studies of lysosomal beta-galactosidase and the non-lysosomal beta-galactosidase-like protein. Biochim Biophys Acta 1999;1455:85–103.

69 Hiraiwa M: Cathepsin A/protective protein: An unusual lysosomal multifunctional protein. Cell Mol Life Sci 1999;56:894–907.
70 Pshezhetsky AV, Potier M: Association of N-acetylgalactosamine-6-sulfate sulfatase with the multienzyme lysosomal complex of beta-galactosidase, cathepsin A, and neuraminidase. Possible implication for intralysosomal catabolism of keratan sulfate. J Biol Chem 1996;271:28359–28365.
71 Dell'Angelica EC, Mullins C, Caplan S, Bonifacino JS: Lysosome-related organelles. FASEB J 2000;14:1265–1278.
72 Kjeldsen L, Calafat J, Borregaard N: Giant granules of neutrophils in Chediak-Higashi syndrome are derived from azurophil granules but not from specific and gelatinase granules. J Leukoc Biol 1998;64:72–77.
73 Singh I: Biochemistry of peroxisomes in health and disease. Mol Cell Biochem 1997;167:1–29.
74 Dansen TB, Wirtz KW: The peroxisome in oxidative stress. IUBMB Life 2001;51:223–230.
75 Baumgartner MR, Saudubray JM: Peroxisomal disorders. Semin Neonatol 2002;7:85–94.
76 Gartner J: Organelle disease: Peroxisomal disorders. Eur J Pediatr 2000;159(Suppl 3):S236–S239.
77 Lazarow PB: Peroxisome biogenesis: Advances and conundrums. Curr Opin Cell Biol 2003; 15:489–497.
78 Brown LA, Baker A: Peroxisome biogenesis and the role of protein import. J Cell Mol Med 2003;7:388–400.
79 Vizeacoumar FJ, Torres-Guzman JC, Bouard D, Aitchison JD, Rachubinski RA: Pex30p, Pex31p and Pex32p form a family of peroxisomal integral membrane proteins regulating peroxisome size and number in *Saccharomyces cervisiae.* Mol Biol Cell 2004;15:665–677.
80 Eckert JH, Erdmann R: Peroxisome biogenesis. Rev Physiol Biochem Pharmacol 2003; 147:75–121.
81 Subramani S: Convergence of model systems for peroxisome biogenesis. Curr Opin Cell Biol 1996;8:513–518.
82 Suzuki Y, Shimozawa N, Imamura A, Fukuda S, Zhang Z, Orii T, Kondo N: Clinical, biochemical and genetic aspects and neuronal migration in peroxisome biogenesis disorders. J Inherit Metab Dis 2001;24:151–165.
83 Birschman I, Stroobants A, van den Berg M, Schafer A, Rosenkranz K, Kunau WH, Tabak H: Pex15p of *Saccharomyces cerevisiae* provides a molecular basis for recruitment of the AAA peroxin Pex6p to peroxisomal membranes. Mol Biol Cell 2003;14:2226–2236.
84 Purdue PE, Lazarow PB: Peroxisome biogenesis. Annu Rev Cell Dev Biol 2001;17:701–752.
85 Li X, Baumgart E, Dong GX, Morrell JC, Jimenez-Sanchez G, Valle D, Smith KD, Gould SJ: PEX11alpha is required for peroxisome proliferation in response to 4-phenylbutyrate but is dispensable for peroxisome proliferator-activated receptor alpha-mediated peroxisome proliferation. Mol Cell Biol 2002;22:8226–8240.
86 Gould SJ, Valle D: Peroxisome biogenesis disorders: Genetics and cell biology. Trends Genet 2000;16:340–345.
87 Gootjes J, Mooijer PA, Dekker C, Barth PG, Poll-The BT, Waterham HR, Wanders RJ: Biochemical markers predicting survival in peroxisome biogenesis disorders. Neurology 2002;59:1746–1749.
88 Preuss N, Brosius U, Biermanns M, Muntau AC, Conzelmann E, Gartner J: PEX1 mutations in complementation group 1 of Zellweger spectrum patients correlate with severity of disease. Pediatr Res 2002;51:706–714.
89 Maxwell MA, Nelson PV, Chin SJ, Paton BC, Carey WF, Crane DI: A common PEX1 frameshift mutation in patients with disorders of peroxisome biogenesis correlates with the severe Zellweger syndrome phenotype. Hum Genet 1999;105:38–44.
90 Purdue PE, Zhang JW, Skoneczny M, Lazarow PB: Rhizomelic chondrodysplasia punctata is caused by deficiency of human PEX7, a homologue of the yeast PTS2 receptor. Nat Genet 1997;15:381–384.
91 Purdue PE, Skoneczny M, Yang X, Zhang JW, Lazarow PB: Rhizomelic chondrodysplasia punctata, a peroxisomal biogenesis disorder caused by defects in Pex7p, a peroxisomal protein import receptor: A minireview. Neurochem Res1999;24:581–586.
92 Danpure CJ: Primary hyperoxaluria; in Scriver CR, et al (eds): The Metabolic and Molecular Bases of Inherited Disease. New York, McGraw-Hill, 2001, pp 3323–3367.

Review

Medical Principles and Practice

Med Princ Pract 2005;14(suppl 1):58–68
DOI: 10.1159/000086185

Received: May 10, 2004
Revised: August 14, 2004

Disorders of Androgen Synthesis – from Cholesterol to Dehydroepiandrosterone

Walter L. Miller

Department of Pediatrics, University of California, San Francisco, Calif., USA

Key Words
P450c17 · Dehydroepiandrosterone · Polycystic ovary syndrome · Adrenarche · Steroidogenic acute regulatory protein · Steroids · P450 oxidoreductase · Hormones · Antley-Bixler syndrome · Congenital lipoid adrenal hyperplasia

Abstract
Androgens and estrogens are primarily made from dehydroepiandrosterone (DHEA), which is made from cholesterol via four steps. First, cholesterol enters the mitochondria with the assistance of the steroidogenic acute regulatory protein (StAR). Mutations in the StAR gene cause congenital lipoid adrenal hyperplasia (lipoid CAH), a potentially lethal disease in which virtually no steroids are made. Lipoid CAH is common among Palestinian Arabs and people from eastern Arabia, and among Korean and Japanese people. Second, within the mitochondria, cholesterol is converted to pregnenolone by the cholesterol side chain cleavage enzyme, P450scc; disorder of this enzyme is very rare, probably due to embryonic lethality. Third, pregnenolone undergoes 17α-hydroxylation by microsomal P450c17. 17α-Hydroxylase deficiency, manifesting as female sexual infantilism and hypertension, is rare except in Brazil. Finally, 17-OH pregnenolone is converted to DHEA by the 17,20 lyase activity of P450c17. The ratio of the 17,20 lyase to 17α-hydroxylase activity of P450c17 determines the ratio of C21 to C19 steroids produced. This ratio is regulated posttranslationally by at least three factors: the abundance of the electron-donating protein P450 oxidoreductase (POR), the presence of cytochrome b_5 and the serine phosphorylation of P450c17. Mutations of POR are a new, recently described disorder manifesting as the Antley-Bixler skeletal dysplasia syndrome, and a form of polycystic ovary syndrome.

Steroid Hormones

Steroid hormones are essential for life and reproduction. There are several classes of steroid hormones, defined according to their physiologic actions and the receptors to which they bind. Mineralocorticoids instruct the kidney to retain sodium. In the absence of mineralocorticoids, the kidney discards Na^+ and retains K^+ and H^+, leading to hyponatremia, hyperkalemia, acidosis, dehydration and death. Thus, mineralocorticoids play a central role in salt and water metabolism and in regulating blood pressure [1]. Glucocorticoids help to maintain euglycemia, mediate stress responses and regulate a variety of immune responses. Survival is possible in the absence of glucocorticoids (Addison's disease) until the time of severe physiologic stress (febrile illness, major surgery, hypovolemia) [2]. Three distinct classes of sex steroids are required for sexual differentiation and repro-

KARGER
Fax +41 61 306 12 34
E-Mail karger@karger.ch
www.karger.com

1011–7571/05/0147–0058$22.00/0

Accessible online at:
www.karger.com/mpp

Prof. Walter L. Miller
Department of Pediatrics, Bldg. MR IV, Room 209
University of California
San Francisco, CA 94142-0978 (USA)
Tel. +1 415 476 2598, Fax +1 415 476 6286, E-Mail wlmlab@itsa.ucsf.edu

duction: *progestins* are required for normal menstrual cycles and for the maintenance of pregnancy; *estrogens* for female sexual development and reproduction; and *androgens* for male sexual differentiation in fetal life, male sexual development, and reproduction. While the individual can survive without sex steroids, their loss would eventually be lethal to the species. Thus, steroid hormones play a central, crucial and irreplaceable role in human life.

Mitochondria, Steroidogenic Acute Regulatory Protein and P450scc

The first, rate-limiting and hormonally regulated step in the biosynthesis of all steroid hormones is the conversion of cholesterol to pregnenolone, described in the next paragraph. Steroidogenic cells do not store significant quantities of steroid hormones; hence steroid secretion is directly related to steroid synthesis. Steroid synthesis is regulated both acutely and chronically. The acute stimulation of steroidogenesis, as in the rapid rise in cortisol in response to intravenous administration of ACTH, is mediated at the level of cholesterol import into mitochondria, which is facilitated by the steroidogenic acute regulatory protein (StAR) [3, 4]. Chronic stimulation, over a period of hours to weeks, such as in the hypercortisolism of Cushing's disease, is mediated by increased transcription of the genes encoding the various steroidogenic enzymes. Thus, the acutely regulated step is the availability of substrate to, and the chronically regulated step is the amount of, those enzymes.

The conversion of cholesterol to pregnenolone is accomplished by the cleavage of the cholesterol side chain, catalyzed by a mitochondrial cytochrome P450 enzyme termed P450scc, where scc designates side chain cleavage. This single enzyme, encoded by a single gene [5, 6] (sometimes termed CYP11A) on chromosome 15q23–q24 [7], catalyzes three distinct sequential reactions on a single active site: cholesterol sequentially undergoes 20-hydroxylation, 22-hydroxylation, and scission of the 20,22 carbon-carbon bond to yield pregnenolone and isocaproaldehyde [8]. The expression of the gene for P450scc enables a cell to become steroidogenic (able to make steroids). The level of P450scc gene transcription is hormonally regulated by multiple pathways, permitting independent regulation of steroidogenesis in multiple cell types [9–11], and the amount of P450scc mRNA present in a cell directly correlates with its steroidogenic capacity [12, 13]. Each of the three reactions catalyzed by P450scc requires a pair of electrons. A flavoprotein (ferredoxin reductase) receives the electrons from NADPH, then passes them to an iron/sulfur protein (ferredoxin), which in turn passes them to the P450. Ferredoxin reductase and ferredoxin are generic electron transfer proteins that can donate electrons to any of the various mitochondrial P450 enzymes [14–17]. P450scc can only function within the mitochondria [18]; hence, delivery of cholesterol to the inner mitochondrial membrane by StAR is a crucial step in steroidogenesis.

Early work showed that ACTH could induce adrenal steroidogenesis very rapidly, and that this induction could be inhibited by cycloheximide or other inhibitors of protein synthesis [19–22], suggesting that a short-lived protein was an essential trigger of the acute response. The responsible protein, StAR, was first identified in rat adrenals and corpus luteum as a phosphoprotein, seen on 2-dimensional gels that appeared rapidly when cells were stimulated with cAMP [23–25]. Stocco's laboratory then cloned the mouse cDNA and found that expression of StAR in mouse Leydig MA-10 cells increased steroidogenesis 6-fold [26]. StAR similarly increased steroidogenesis in COS-1 cells cotransfected with the P450scc system [27, 28], but the definitive proof that StAR is essential in human physiology came from finding StAR mutations in congenital lipoid adrenal hyperplasia (lipoid CAH) [27, 29].

Congenital Lipoid Adrenal Hyperplasia

Lipoid CAH is a rare autosomal recessive disorder that severely disrupts the synthesis of all adrenal and gonadal steroids [4]. A severe defect in fetal testicular biosynthesis is evident because affected 46,XY genetic males are born with wholly female external genitalia, reflecting an absence of testosterone synthesis between 6 and 12 weeks of gestation. At birth, the adrenals are engorged with cholesterol ester deposits. Affected newborns have low but measurable levels of steroid hormones. They soon die from glucocorticoid and mineralocorticoid deficiency if hormonal replacement therapy is not initiated, but properly treated patients can survive to adulthood [30, 31].

Although StAR is needed for a rapid and maximal steroidogenic response, there are also low levels of StAR-independent steroidogenesis in cells that express StAR [27, 32]. However, some steroidogenic cells, such as those of the placenta [28], do not express StAR. The brain also expresses low levels of StAR in apparently steroidogenic cells that also contain P450scc [33–35], but the role of StAR in the brain is not clear, as there is no known tropic

Table 1. Lipoid CAH in Palestinians

Nucleotide	Amino acid	Families	Alleles	Countries of origin
A632G	E169G	1	1	Israel
G671T	R182L	4	7	Denmark, Jordan
ΔC650	frameshift	1	2	Kuwait
Δ2T593	frameshift	2	3	Jordan
C703T	R193X	1	2	Israel

Seven patients have been reported from 6 families, 1 of which was consanguineous, thus representing 11 unique alleles. Several alleles carried more than 1 mutation, so that the total number of affected alleles is greater than 11. Data are from Bose et al. [32, 43].

stimulus to neurosteroidogenesis [36], and a specific CNS defect in StAR-deficient patients with lipoid CAH has not been reported [36]. The demonstration of StAR-independent steroidogenesis led to formulation of the two-hit model of lipoid CAH [32]. The first hit is the mutation in the StAR gene, ablating StAR-dependent steroidogenesis but permitting StAR-independent steroidogenesis to persist. StAR-independent placental steroidogenesis and placental synthesis of progesterone remain normal, permitting term gestation. Low levels of adrenal StAR-independent steroidogenesis account for the low, but detectable, levels of steroid hormones seen in the sera of lipoid CAH patients in the first month of life [31, 32]. The presence of these steroids appears to explain why untreated lipoid CAH infants may survive without treatment for several months [31, 32, 37, 45], whereas patients with other forms of salt-wasting CAH do not. However, these steroid hormone concentrations are too low to suppress secretion of ACTH, gonadotropins and angiotensin II to normal levels. Excess amounts of these tropic hormones stimulate cellular uptake of low-density lipoprotein cholesterol and increased production of cholesterol from acetate. This tropic hormone stimulation results in the accumulation of cholesterol esters, which eventually disrupts the cells, either via physical engorgement of the cell with droplets of cholesterol esters or by a chemical action of cholesterol oxidation products, or both. This second hit thus destroys the low levels of StAR-independent steroidogenesis, leading to the unmeasurable levels of steroid in the serum of older children with lipoid CAH, and to the absence of circulating testosterone in affected 46,XY fetuses [4, 32].

The fetal ovaries do not express the genes for the steroidogenic enzymes and thus do not make steroids [38]; unlike the testes and adrenals, the ovaries only start to make steroid hormones at the onset of puberty. Consequently, the ovaries of 46,XX females affected with lipoid CAH do not receive the second hit until the onset of puberty, when LH stimulates low levels of StAR-independent steroidogenesis. With each monthly cycle, another follicle is recruited and stimulated by gonadotropins, producing spontaneous age-appropriate breast development in affected 46,XX individuals. However, gonadotropin stimulation quickly results in cholesterol engorgement of these cells (the second hit in lipoid CAH) so that the later phase of ovarian steroidogenesis, the secretion of large amounts of progesterone, does not occur [39, 40]. Follicles that are not recruited remain unstimulated and constitute a reservoir of steroidogenic cells undamaged by the second hit of lipoid CAH, so that with each monthly cycle a new undamaged follicle is recruited. Estrogen is then produced, leading to cyclic uterine estrogen withdrawal bleeding that resembles normal menses, but, as there is no progesterone, these cycles are anovulatory. The predictions of the two-hit model have recently been confirmed by elegant experiments with StAR knockout mice [41, 42].

StAR mutations causing lipoid CAH have been described in about 70 patients from all over the world. Lipoid CAH is most common in Japan and Korea [27, 32, 37, 43, 44]. The mutation Q258X accounts for about 70% of affected alleles in Japan and 95% of the alleles reported to date from Korea. A second population in whom lipoid CAH appears to be common is Palestinian Arabs, among whom 7 cases have been reported to date. However, several different mutations are found in these patients, suggesting founder effects in limited gene pools, and consanguinity (table 1). A third group in whom lipoid CAH appears to be common is Arabs from the Eastern Province of Saudi Arabia and nearby Qatar, all of whom carry the mutation R182H, suggesting a founder effect [45].

True P450scc Deficiency

When lipoid CAH was first described, long before the discovery of StAR, it was generally termed '20,22 desmolase deficiency' because it was thought to be a disorder in the enzyme (P450scc) that converts cholesterol to pregnenolone [31]. However, analysis of the gene for P450scc was normal in these patients [46, 47]. Finding StAR mutations in virtually all patients with the lipoid CAH phenotype thus posed the question: 'Why do we not see mutations in P450scc, as we do in all other steroidogenic enzymes?' As P450scc is required for placental steroidogenesis, and placental progesterone is required to suppress uterine contractility to maintain pregnancy, we inferred that absence of P450scc and placental steroidogenesis would lead to spontaneous abortion, so that P450scc deficiency would result in embryonic lethality [48]. This view appears to be consistent with two recent reports of the lipoid CAH phenotype in patients with partial deficiency of P450scc. One patient carried a de novo heterozygous mutation not found in either parent [49]; the other was a compound heterozygote, but one of the 'mutations' was simply a polymorphism having no impact on P450scc catalysis [50]. In these cases, the lipoid CAH phenotype can again be attributed to a modification of the two-hit model of lipoid CAH [49]. Thus, unusual patients provide definitive tests of models formulated from observations from experiments in vitro.

The Biochemistry of StAR

StAR is a most unusual protein. It is initially synthesized as a 37-kDa protein that is cleaved to a 30-kDa form when it is imported into mitochondria. Cloning studies and analyses of DNA databases show that StAR is the first member of a new class of proteins that have so-called START (StAR-related lipid transfer) domains, which are structurally related to residues 62–285 of StAR [51–53]. The X-ray crystal structure of two of these proteins, N-218 MLN64 [54] and StarD4 [55], has been determined, showing that each has a helix-grip fold encompassing a hydrophobic pocket that appears to be able to bind a single molecule of cholesterol. While StAR/cholesterol binding curves also indicate that StAR binds a single cholesterol molecule [54], experiments with StAR-induced lipid transfer between liposomes show that each molecule of StAR can transfer several cholesterol molecules in vitro [56, 57] and experiments in vivo suggest that each molecule of StAR can promote the movement of hundreds of molecules of cholesterol from the outer mitochondrial membrane (OMM) to the inner mitochondrial membrane (IMM) [58]. This fostered the view that StAR functions in the mitochondrial intramembranous space (IMS) to shuttle cholesterol from OMM to IMM. Furthermore, the cleavage of the 37-kDa StAR to the 30-kDa StAR, similarly to other mitochondrial proteins, led to the mistaken notion that the 37-kDa form is a 'precursor' and the 30-kDa protein is the 'mature' form. However, it is now clear that StAR's activity is determined by its cellular location, and not by its proteolytic cleavage status. Hence the 37-kDa form is 'mature' (i.e. active) and the 30-kDa form represents the first step in the intramitochondrial degradation of StAR [59]. Deleting the mitochondrial leader peptide prevents StAR from entering the mitochondria and confines it to the cytoplasm; yet this N-62 StAR molecule remains fully active [60], and is also able to transfer cholesterol to other membranes, such as the endoplasmic reticulum [61]. Thus, it appeared that StAR's mitochondrial leader peptide directed StAR's activity to the mitochondria and was confined to that organelle, preventing it from transferring cholesterol to other membranes, and that StAR's activity was confined to the OMM [62]. The amino-terminal domain of StAR (residues 63–193) is protease resistant, suggesting tight protein folding [63]. This would presumably slow mitochondrial import, prolonging contact with the OMM, and increasing activity. N-62 StAR undergoes a conformational change in acidic environments, both in solution and in synthetic membranes [63, 64]. This conformational change appears to foster association with the OMM, facilitating binding of OMM cholesterol for transfer to the IMM. However, until recently this view had been controversial.

Using novel fusion proteins that affix StAR to the OMM, confine it to the IMS, or affix it to the IMM, we have now proven that StAR acts exclusively on the OMM, and is wholly inactive in all other mitochondrial compartments [65]. By fusing StAR to a protein (Tom 20) that is a component of the OMM, we confined StAR to the cytoplasmic side of the OMM and prevented its mitochondrial entry, resulting in a dramatic increase in activity. This suggests that the level of StAR activity is proportional to the time it resides on the OMM. To test this, we built variants of StAR with altered mitochondrial leaders designed to either slow or speed its mitochondrial entry. Constructs designed for slow entry increased activity both in intact cells and in isolated mitochondria; constructs designed to speed entry similarly decreased activity. Using an in vitro cell-free transcription/translation system and in vitro assays of the kinetics of mitochondrial protein import, we could show that each construct either increased or decreased the speed of mitochondrial protein entry, as designed, and that the level of StAR activity was inversely proportional to its speed of entry [65]. Recent experiments have shown that only the carboxyl-terminal α-helix of StAR interacts with the OMM [65a]. Thus, StAR acts exclusively on the OMM to facilitate the transfer of many molecules of cholesterol to the IMM. StAR then is imported into the mitochondria and cleaved to the 30-kDa form. This 30-kDa StAR protein is inactive, not because of the proteolytic cleavage but because of its location in the mitochondrial matrix from where it cannot reach the cytoplasmic aspect of the OMM [65].

P450c17 – The Qualitative Regulator of Steroidogenesis

Whereas the StAR/P450scc system is the rate-limiting and hormonally regulated step, and hence is the quantitative regulator of steroidogenesis, P450c17 is the qualitative regulator that determines which class of steroids that would be produced [8]. When P450c17 is absent, such as in the adrenal zona glomerulosa, the products are C21 (21-carbon) 17-deoxysteroids, such as aldosterone. When the 17α-hydroxylase and 17,20 lyase activities of P450c17 are present, C19 precursors of sex steroids are produced. Because all steroid hormones are made from the pregnenolone produced by P450scc, the presence or absence of each of the activities of P450c17 directs this pregnenolone toward its final metabolic pathway. While all cytochrome P450 enzymes can catalyze multiple reactions on a single active site, P450c17 is the only one described to date in which these multiple activities are differentially regulated by a physiologic process. Unlike P450scc, which is found in mitochondria, P450c17 is found in the endoplasmic reticulum. These 'microsomal' P450 enzymes, which include all of the drug-metabolizing hepatic cytochrome P450 enzymes, receive electrons from NADPH via a flavoprotein intermediate called P450 oxidoreductase (POR).

P450c17 – One Enzyme with Two Activities

Until the 1980s, observations of human physiology led to the mistaken belief that 17α-hydroxylase and 17,20 lyase were separate enzymes, regulated independently. The adrenals of young children make very little dehydroepiandrosterone (DHEA) until the onset of adrenarche and the development of the adrenal zona reticularis. Andrenarche begins at about age 6–8, is independent of the gonads or gonadotropins, and continues after the end of puberty, since serum DHEA and DHEA-S concentrations are highest between ages 25 and 30 [66–68]. This profound increase in adrenal DHEA synthesis proceeds without changes in the serum concentrations of cortisol and ACTH [66], and thus appears to involve selective induction of 17,20 lyase activity, while 17α-hydroxylase activity, as indicated by the age-independent cortisol concentrations, remains essentially unchanged. This phenomenon, adrenarche, is difficult to study, because it only occurs in large old-world primates [69, 70]. Searches for a hormonal trigger for adrenarche have been unsuccessful [71–73]. Early reports of familial, apparently autosomal recessive isolated 17,20 lyase deficiency, in which the patients had normal cortisol values and hence normal 17α-hydroxylase activity [74–76], appeared to confirm that 17α-hydroxylase and 17,20 lyase were separate enzymes encoded by separate genes. However, biochemical studies with the pig [77–80] and guinea pig [81, 82] showed that a single protein, cytochrome P450c17, catalyzed both reactions on single active site, and nonsteroidogenic cells transfected with vectors that expressed bovine [83] or human [84] P450c17 acquired both 17α-hydroxylase and 17,20 lyase activities. Accordingly, genetic studies showed that there was a single species of P450c17 mRNA that was identical in the human adrenal and gonad [85], because it was encoded by a single-copy gene [86] (now termed CYP17) on chromosome 10q24.3 [7, 87, 88]. Thus, although the transcriptional regulation of human P450c17 is of central importance [89–91], to regulate the amount of P450c17 present, the differential regulation of the two activities of P450c17 must lie at one or more points downstream from its gene. All present data now indicate that this regulation is mediated by several factors, all of which influence electron flow from NADPH via POR to P450c17 [92].

Role of Redox Partners in Regulating 17,20 Lyase Activity

Hall was first to note that POR is much more abundant in the porcine testis, which has a high ratio of lyase to hydroxylase activity, than in the porcine adrenal, which has low lyase activity [93, 94]. Combining these proteins in vitro showed that POR was needed for both activities, and had suggested that cytochrome b_5 could selectively augment the 17,20 lyase reaction [93–95]. It had generally been thought that cytochrome b_5 can serve as an alternative electron transfer protein, taking the place of POR in certain circumstances [96, 97]. Transfection of nonsteroidogenic monkey kidney COS-1 cells with a vector expressing human P450c17 permits the cells to catalyze 17,20 lyase activity that converts 17-OH pregnenolone to DHEA, but not 17-OH progesterone to androstenedione [84]. When these cells are cotransfected with a vector expressing human POR, conversion of 17-OH pregnenolone to DHEA increases, but 17-OH progesterone is not converted to androstenedione [98]. Thus, qualitative data obtained in whole living cells indicated that human P450c17 catalyzes 17,20 lyase activity only with Δ^5 substrates, and that this activity could be augmented by excess POR.

To understand the enzymology of P450c17 in detail, we utilized yeast systems in which the endogenous yeast genes for POR or b_5 were disrupted, permitting rigorous kinetic analysis of the role of each potential redox partner on each of the four potential reactions of human P450c17 [99]. These data confirm that POR is required for both 17α-hydroxylation and 17,20 lyase activity and that the 17,20 lyase reaction is much more severely impeded by low concentrations of POR than is the 17α-hydroxylase reaction. Cytochrome b_5 selectively augments 17,20 lyase activity when POR is present, but b_5 alone cannot support catalysis. Apo-b_5, which is the cytochrome protein without its heme group, is as effective at promoting 17,20 lyase activity as holo-b_5 which has the heme group, indicating that b_5 augments the 17,20 lyase activity of P450c17 as an allosteric facilitator, and not as an electron donor. The expression of b_5 in the adrenal is largely confined to the zona reticularis [100, 101] and this expression increases during adrenarche [102], so that adrenal levels of b_5 correlate with DHEA secretion. Surprisingly, 17-OH progesterone is a poor substrate for human P450c17. The K_m of the 17,20 lyase reaction is 10 times higher, and the V_{max} is reduced to 30% when Δ^4 17-OH progesterone is the substrate compared to the values with Δ^5 17-OH pregnenolone. Thus, the 17,20 lyase action of human P450c17 is about 30-fold greater with Δ^5 substrates than with Δ^4 substrates. As a result, most androstenedione is produced by the action of 3β-hydroxysteroid dehydrogenase on DHEA, and only minimal quantities derive from 17-OH progesterone, so that most human sex steroid production is from DHEA.

Isolated 17,20 Lyase Deficiency

Several patients with apparent isolated 17,20 lyase deficiency were described in the 1970s and 1980s, but the demonstration that a single enzyme encoded by a single gene catalyzed both 17α-hydroxylase and 17,20 lyase activities put the existence of such a disease in doubt. Further questions were raised when Waterman's group showed that a patient with apparent isolated 17,20 lyase deficiency in fact had P450c17 mutations that caused complete 17α-hydroxylase deficiency [103]. No genetically and enzymologically proven cases of isolated 17,20 lyase deficiency had been reported until 1997, when we studied two patients from rural Brazil who had clinical and hormonal findings consistent with isolated 17,20 lyase deficiency [104].

DNA sequencing showed that each was homozygous for a different mutation; in one patient, the arginine at position 347 was replaced with histidine (R347H) and, in the other patient, the arginine at position 358 was replaced with glutamine (R358Q). Transfection of COS-1 cells with vectors expressing these P450c17 mutants showed that each retained only about 5% of normal 17,20 lyase activity but retained about 65% of normal 17α-hydroxylase activity. Despite this loss of lyase activity, 17-OH pregnenolone acted as a strictly competitive inhibitor of the hydroxylase reaction, showing that these mutations did not affect the active site [105]. However, the 17,20 lyase activity of these mutants could be partially restored by adding excess b_5, suggesting that they affect binding of redox partners and consequent electron transfer [105]. Molecular modeling of P450c17 based on the known X-ray crystallographic structures of several bacterial cytochrome P450 enzymes showed that the two mutations altered the distribution of electrostatic charge on the surface of the P450c17 protein in the region that interacts with POR [106]. Thus, changing a single positive charge to neutral or to a negative charge partially decreased the ability of P450c17 to interact with POR, so that 17α-hydroxylation was relatively unaffected, but 17,20 lyase activity was severely disrupted, in fashion very analogous to the persistence of 17α-hydroxylase activity but the relative absence of 17,20 lyase activity when the concentrations of POR are low. Thus, isolated 17,20 lyase deficiency is an important 'site-directed mutagenesis experiment of nature' that identifies redox partner interactions as crucial regulators of 17,20 lyase activity.

POR Deficiency

A rare variant of 17,20 lyase deficiency seen in association with partial deficiency of steroid 21-hydroxylase was first described in 1985 by Peterson et al. [107] and is now understood in detail [108]. Mutations in the genes for P450c17 and P450c21 (steroid 21-hydroxylase) were sought but not found. In 1986, we suggested that this disorder was a deficiency of POR [109], but this seemed unlikely as POR knockout mice do not survive past the middle of fetal development [110, 111]. Furthermore, most children with apparent combined 17α-hydroxylase and 21-hydroxylase deficiencies also have the Antley-Bixler skeletal malformation syndrome (ABS), consisting of craniosynostosis, radioulnar or radiohumeral synostosis, and mid-face hypoplasia, and half also have genital anomalies. About half of the children with ABS have mutations in fibroblast growth factor receptor 2 (FGFR2), as do most patients with the Apert, Crouzon, Jackson-Weiss and Pfeiffer craniosynostosis syndromes, but the ABS patients who have FGFR2 mutations do not have abnormal steroidogenesis, and those who lack FGFR2 mutations have abnormal steroidogenesis [112]. The availability of the human POR gene sequence provided by the human genome project permitted the examination of the POR gene in this disorder. We found POR mutations in 3 children with ABS and abnormal steroidogenesis, and also in an adult woman with amenorrhea, hypertension, cystic ovaries and hormonal findings suggesting partial combined 17α-hydroxylase and 21-hydroxylase deficiencies [113]. We found five different POR missense mutations, and tested the ability of each to receive electrons from NADPH and to donate them to both a test substrate (cytochrome c) and a real substrate (P450c17). There was substantially better phenotype/genotype correlation with the P450c17 assays than with the cytochrome c assays [113]. This indicates that POR mutants will donate electrons to different substrates with different efficiencies; thus, a POR mutant may severely affect some P450 enzymes but not others. As POR donates electrons to hepatic drug-metabolizing P450 enzymes, minor POR sequence variants may contribute to genetic variations in drug metabolism. The similarity between the hormonal phenotypes in POR deficiency and mutations in the POR (redox partner) binding site of P450c17 further underscores the crucial role of electron donation in regulating 17,20 lyase activity and hence regulating androgen synthesis. We recently reported a large series of 32 such patients, confirming and extending these observations [113a].

Serine Phosphorylation of P450 Increases 17,20 Lyase Activity

Radiolabeling studies of COS-1 cells transfected with a human P450c17 expression vector or of untransfected human adrenal NCI-H295 cells [114], which express all human steroidogenic enzymes [115], showed that human P450c17 is a phosphoprotein [116]. This phosphorylation is rapidly induced by cAMP, occurs on serine and threonine but not on tyrosine residues, and increases the 17,20 lyase activity of P450c17. Conversely, dephosphorylation of microsomes containing P450c17 ablates 17,20 lyase activity without reducing 17α-hydroxylase activity and without affecting steroid binding to the enzyme. The mechanism by which serine phosphorylation promotes 17,20 lyase activity is not yet known. Phosphorylated serine residues near the redox partner binding site may increase the affinity of P450c17 for POR or b_5 (or both), thus effectively increasing the flow of electrons to the P450. The kinase responsible for P450c17 serine phosphorylation has not been identified yet, but recent data indicate that the degree of phosphorylation in vivo is regulated by the counterbalancing actions of the kinase and protein phosphatase 2A, which in turn is regulated by phosphoprotein SET [117].

Serine Phosphorylation, Adrenarche and the Polycystic Ovary Syndrome

The polycystic ovary syndrome (PCOS) is a heterogenous disorder characterized by hirsutism, virilism, hyperandrogenism, menstrual irregularities, chronic anovulation, obesity, insulin resistance, acanthosis nigricans, high concentrations of LH and ovarian cysts. Hyperandrogenism and insulin resistance appear to be primary lesions, and the other findings are secondary events [118]. The hyperandrogenism in women with PCOS is of both ovarian and adrenal origin [119–122]. The adrenal hyperandrogenism of PCOS resembles an exaggerated form of adrenarche, and girls with premature adrenarche are more likely to develop PCOS [123, 124]. A gain-of-function disorder in the pathway leading to the serine/threonine phosphorylation of P450c17 could account for such increases in both adrenal and ovarian androgen secretion and an earlier age of adrenarche [116], but such mutations have not yet been reported. The hyperinsulism and insulin resistance of PCOS is at the level of insulin receptor signal transduction [125, 126]. Serine phosphorylation of the β chain of the insulin receptor interferes with the tyrosine phosphorylation of the receptor that normally follows binding of insulin [127–129]. Furthermore, some PCOS women appear to have insulin receptors in their fibroblasts that are hyperphosphorylated [130]. Thus, a gain-of-function mutation in a serine-threonine kinase or its signal transduction pathway might increase the serine hyperphosphorylation of both P450c17 and the β chain of the insulin receptor, thus accounting for both the hyperandrogenism and insulin resistance of PCOS with a single molecular lesion.

Conclusions

Defects in the synthesis of DHEA cause deficiencies in the synthesis of androgens and estrogens. In the first stage, mutations in the StAR gene cause potentially lethal lipoid CAH. Defects in the cholesterol side chain cleavage enzyme, the second step, are rare. Defects in the third stage, 17-hydroxylation lead to hypertension and female sexual infantilism. Defects in the final stage of DHEA synthesis, 17,20 lyase activity, are due to disorders of electron transfer to P450c17, including mutations in POR. Of these, the defect in StAR is common among Palestinian Arabs and in Eastern Arabia.

References

1 Fardella CE, Miller WL: Molecular biology of mineralocorticoid metabolism. Ann Rev Nutrition 1996;16:443–470.
2 Miller WL, Chrousos GP: The adrenal cortex; in Felig P, Frohman L (eds): Endocrinology and Metabolism, ed 4. New York, McGraw-Hill, 2001, pp 387–524.
3 Stocco DM, Clark BJ: Regulation of the acute production of steroids in steroidogenic cells. Endocr Rev 1996;17:221–244.
4 Miller WL: Congenital lipoid adrenal hyperplasia: The human knockout of the steroidogenic acute regulatory protein. J Mol Endocrinol 1997;19:227–240.
5 Chung B, Matteson KJ, Voutilainen R, Mohandas TK, Miller WL: Human cholesterol side-chain cleavage enzyme, P450scc: cDNA cloning, assignment of the gene to chromosome 15, and expression in the placenta. Proc Natl Acad Sci USA 1986;83:8962–8966.
6 Morohashi K, Sogawa K, Omura T, Fujii-Kuriyama Y: Gene structure of human cytochrome P-450(scc), cholesterol desmolase. J Biochem 1987;101:879–887.
7 Sparkes RS, Klisak I, Miller WL: Regional mapping of genes encoding human steroidogenic enzymes: P450scc to 15q23-q24, adrenodoxin to 11q22; adrenodoxin reductase to 17q24-q25; and P450c17 to 10q24-q25. DNA Cell Biol 1991;10:359–365.
8 Miller WL: Molecular biology of steroid hormone synthesis. Endocr Rev 1988;9:295–318.
9 Moore CCD, Brentano ST, Miller WL: Human P450scc gene transcription is induced through cyclic AMP and repressed by 12-*O*-tetradecanoylphorbol-13-acetate and A23187 by independent *cis*-elements. Mol Cell Biol 1990;10:6013–6023.
10 Moore CCD, Hum DW, Miller WL: Identification of positive and negative placental-specific basal elements and a cyclic adenosine 3′,5′-monophosphate response element in the human gene for P450scc. Mol Endocrinol 1992;6:2045–2058.
11 Huang N, Miller WL: Cloning of factors related to HIV-inducible LBP proteins that regulate steroidogenic factor-1-independent human placental expression of the cholesterol side-chain cleavage enzyme, P450scc. J Biol Chem 2000;275:2852–2858.
12 Voutilainen R, Tapanainen J, Chung B, Matteson KJ, Miller WL: Hormonal regulation of P450scc (20,22 desmolase) and P450c17 (17α-hydroxylase/17,20 lyase) in cultured human granulosa cells. J Clin Endocrinol Metab 1986;63:202–207.
13 DiBlasio AM, Voutilainen R, Jaffe RB, Miller WL: Hormonal regulation of mRNA for P450scc (cholesterol side-chain cleavage enzyme) and P450c17 (17α-hydroxylase/17,20 lyase) in cultured human fetal adrenal cells. J Clin Endocrinol Metab 1987;65:170–175.
14 Picado-Leonard J, Voutilainen R, Kao L, Chung B, Strauss JF 3rd, Miller WL: Human adrenodoxin: Cloning of three cDNAs and cycloheximide enhancement in JEG-3 cells. J Biol Chem 1988;263:3240–3244.
15 Solish S, Picado-Leonard J, Morel Y, Kuhn RW, Mohandas TK, Hanukoglu I, Miller WL: Human adrenodoxin reductase: Two mRNAs encoded by a single gene on chromosome 17 cen→q25 are expressed in steroidogenic tissues. Proc Natl Acad Sci USA 1988;85:7104–7108.
16 Chang C-Y, Wu D-A, Lai C-C, Miller WL, Chung B: Cloning and structure of the human adrenodoxin gene. DNA 1988;7:609–615.
17 Lin D, Shi Y, Miller WL: Cloning and sequence of the human adrenodoxin reductase gene. Proc Natl Acad Sci USA 1990;87:8516–8520.
18 Black SM, Harikrishna JA, Szklarz GD, Miller WL: The mitochondrial environment is required for activity of the cholesterol side-chain cleavage enzyme, cytochrome P450scc. Proc Natl Acad Sci USA 1994;91:7247–7251.
19 Stone D, Hechter O: Studies on ACTH action in perfused bovine adrenals: Aspects of progesterone as an intermediary in cortico-steroidogenesis. Arch Biochem Biophys 1955;54:121–128.
20 Ferguson JJ: Protein synthesis and adrenocorticotropin responsiveness. J Biol Chem 1963;238:2754–2759.
21 Garren LD, Ney RL, Davis WW: Studies on the role of protein synthesis in the regulation of corticosterone production by ACTH in vivo. Proc Natl Acad Sci USA 1965;53:1443–1450.
22 Garren LD, Davis WW, Crocco RM: Puromycin analogs: Action of adrenocorticotropic hormone and the rolc of glycogen. Science 1966;152:1386–1388.
23 Pon LA, Orme-Johnson NR: Acute stimulation of steroidogenesis in corpus luteum and adrenal cortex by peptide hormones. J Biol Chem 1986;261:6594–6599.
24 Epstein LF, Orme-Johnson NR: Regulation of steroid hormone biosynthesis. Identification of precursors of a phosphoprotein targeted to the mitochondrion in stimulated rat adrenal cortex cells. J Biol Chem 1991;266:19739–19745.
25 Stocco DM, Sodeman TC: The 30-kDa mitochondrial protein induced by hormone stimulation in MA-10 mouse Leydig tumor cells are processed from larger precursors. J Biol Chem 1991;266:19731–19738.
26 Clark BJ, Wells J, King SR, Stocco DM: The purification, cloning and expression of a novel luteinizing hormone-induced mitochondrial protein in MA-10 mouse Leydig tumor cells. Characterization of the steroidogenic acute regulatory protein (StAR). J Biol Chem 1994;269:28314–28322.
27 Lin D, Sugawara T, Strauss JF 3rd, Clark BJ, Stocco, DM, Saenger P, Rogol A, Miller WL: Role of steroidogenic acute regulatory protein in adrenal and gonadal steroidogenesis. Science 1995;267:1828–1831.
28 Saenger P, Klonari Z, Black SM, Compagnone N, Mellon SH, Fleischer A, Abrams CAL, Shackleton CHL, Miller WL: Prenatal diagnosis of congenital lipoid adrenal hyperplasia. J Clin Endocrinol Metab 1995;80:200–205.
29 Tee MK, Lin D, Sugawara T, Holt JA, Guiguen Y, Buckingham B, Strauss JF 3rd, Miller WL: T→A transversion 11 bp from a splice acceptor site in the gene for steroidogenic acute regulatory protein causes congenital lipoid adrenal hyperplasia. Hum Mol Genet 1995;4:2299–2305.
30 Kirkland RT, Kirkland JL, Johnson CM, Horning MG, Librik L, Clayton GW: Congenital lipoid adrenal hyperplasia in an eight-year-old phenotypic female. J Clin Endocrinol Metab 1973;36:488–496.31.
31 Hauffa BP, Miller WL, Grumbach MM, Conte FA, Kaplan SL: Congenital adrenal hyperplasia due to deficient cholesterol side-chain cleavage activity (20,22 desmolase) in a patient treated for 18 years. Clin Endocrinol 1985;23:481–493.
32 Bose HS, Sugawara T, Strauss JF 3rd, Miller WL: The pathophysiology and genetics of congenital lipoid adrenal hyperplasia. N Engl J Med 1996;335:1870–1878.

33 Furukawa A, Miyatake A, Ohnishi T, Ichikawa Y: Steroidogenic acute regulatory protein (StAR) transcripts constituitively in the adult rat central nervous system: Colocalization of StAR, cytochrome P450scc (CYP XIA1) and 3β-hydroxysteroid dehydrogenase in the rat brain. J Neurochem 1998;71:2231–2238.
34 Kimoto T, Tsurugizawa T, Ohta Y, Makino J, Tamura H, Hojo Y, Takata N, Kawato S: Neurosteroid synthesis by cytochrome p450-containing systems localized in the rat brain hippocampal neurons: N-methyl-*D*-aspartate and calcium-dependent synthesis. Endocrinology 2001;142:3578–3589.
35 King SR, Manna PR, Ishii T, Syapin PJ, Ginsberg SD, Wilson K, Walsh LP, Parker KL, Stocco DM, Smith RG, Lamb DJ: An essential component in steroid synthesis, the steroidogenic acute regulatory protein, is expressed in discrete regions of the brain. J Neurosci 2002; 22:10613–10620.
36 Compagnone NA, Mellon, SH: Neurosteroids: Biosynthesis and function of these novel neuromodulators. Front Neuroendocrinol 2000; 21:1–56.
37 Nakae J, Tajima T, Sugawara T, Arakane F, Hanaki K, Hotsubo T, Igarashi N, Igarashi Y, Ishii T, Koda N, Kondo T, Kohno H, Nakagawa Y, Tachibana K, Takeshima Y, Tsubouchi K, Strauss JF 3rd, Fujieda K: Analysis of the steroidogenic acute regulatory protein (StAR) gene in Japanese patients with congenital lipoid adrenal hyperplasia. Hum Mol Genet 1997;6:571–576.
38 Voutilainen R, Miller WL: Developmental expression of genes for the steroidogenic enzymes P450scc (20,22 desmolase), P450c17 (17α-hydroxylase/17,20 lyase) and P450c21 (21-hydroxylase) in the human fetus. J Clin Endocrinol Metab 1986;63:1145–1150.
39 Bose HS, Pescovitz OH, Miller WL: Spontaneous feminization in a 46,XX female patient with congenital lipoid adrenal hyperplasia caused by a homozygous frame-shift mutation in the steroidogenic acute regulatory protein. J Clin Endocrinol Metab 1997;82:1511–1515.
40 Fujieda K, Tajima T, Nakae J, Sageshima S, Tachibana K, Suwa S, Sugawara T, Strauss JF 3rd: Spontaneous puberty in 46,XX subjects with congenital lipoid adrenal hyperplasia. J Clin Invest 1997;99:1265–1271.
41 Caron KM, Soo S-C, Wetsel WC, Stocco DM, Clark BJ, Parker KL: Targeted disruption of the mouse gene encoding steroidogenic acute regulatory protein provides insights into congenital lipoid adrenal hyperplasia. Proc Natl Acad Sci USA 1997;94:11540–11545.
42 Hasegawa T, Zhao L, Caron KM, Majdic G, Suzuki T, Shizawa S, Sasano H, Parker KL: Developmental roles of the steroidogenic acute regulatory protein (StAR) as revealed by StAR knockout mice. Mol Endocrinol 2000;14: 1462–1471.
43 Bose HS, Sato S, Aisenberg J, Shalev SA, Matsuo N, Miller WL: Mutations in the steroidogenic acute regulatory protein (StAR) in six patients with congenital lipoid adrenal hyperplasia. J Clin Endocrinol Metab 2000;85: 3636–3639.
44 Yoo HW, Kim GH: Molecular and clinical characterization of Korean patients with congenital lipoid adrenal hyperplasia. J Pediatr Endocrinol Metab 1998;11:707–711.
45 Chen X, Baker BY, Abduljabbar MA, Miller WL: A genetic isolate of congenital lipoid adrenal hyperplasia with atypical clinical features. J Clin Endocrinol Metab 2005;90:835–840.
46 Matteson KJ, Chung B, Urdea MS, Miller WL: Study of cholesterol side-chain cleavage (20,22 desmolase) deficiency causing congenital lipoid adrenal hyperplasia using bovine-sequence P450scc oligodeoxyribonucleotide probes. Endocrinology 1986;118:1296–1305.
47 Lin D, Gitelman SE, Saenger P, Miller WL: Normal genes for the cholesterol side chain cleavage enzyme, P450scc, in congenital lipoid adrenal hyperplasia. J Clin Invest 1991;88: 1955–1962.
48 Miller WL: Why nobody has P450scc (20,22 desmolase) deficiency (letter to the editor). J Clin Endocrinol Metab 1998;83:1399–1400.
49 Tajima T, Fujieda K, Kouda N, Nakae J, Miller WL: Heterozygous mutation in the cholesterol side chain cleavage enzyme (P450scc) gene in a patient with 46,XY sex reversal and adrenal insufficiency. J Clin Endocrinol Metab 2001;86:3820–3825.
50 Katsumata N, Ohtake M, Hojo T, Ogawa E, Hara T, Sato N, Tanaka T: Compound heterozygous mutations in the cholesterol side-chain cleavage enzyme gene (CYP11A) cause congenital adrenal insufficiency in humans. J Clin Endocrinol Metab 2002;87:3808–3813.
51 Ponting CP, Aravind L: START: A lipid-binding domain in StAR, HD-ZIP and signaling proteins. Trends Biochem Sci 1999;24:130–132.
52 Iyer LM, Koonin EV, Aravind L: Adaptations of the helix-grip fold for ligand binding and catalysis in the START domain superfamily. Proteins 2001;43:134–144.
53 Soccio RE, Breslow JL: StAR-related lipid transfer (START) proteins: Mediators of intracellular lipid metabolism. J Biol Chem 2003; 278:22183–22186.
54 Tsujishita Y, Hurley JH: Structure and lipid transport mechanism of a StAR-related domain. Nat Struct Biol 2000;7:408–414.
55 Romanowski MJ, Soccio RE, Breslow JL, Burley SK: Crystal structure of the *Mus musculus* cholesterol-regulated START protein 4 (StarD4) containing a StAR-related lipid transfer domain. Proc Natl Acad Sci USA 2002;99: 6949–6954.
56 Tuckey RC, Headlam MJ, Bose HS, Miller WL: Transfer of cholesterol between phospholipid vesicles mediated by the steroidogenic acute regulatory protein (StAR). J Biol Chem 2002;277:47123–47128.
57 Tuckey RC, Bose HS, Czerwionka I, Miller WL: Molten globule structure and steroidogenic activity of N-218 MLN64 in human placental mitochondria. Endocrinology 2004;145: 1700–1707.
58 Artemenko I, Zhao D, Hales DB, Hales KH, Jefcoate CR: Mitochondrial processing of newly synthesized steroidogenic acute regulatory protein (StAR), but not total StAR, mediates cholesterol transfer to cytochrome P450 side chain cleavage enzyme in adrenal cells. J Biol Chem 2001;276:46583–46596.
59 Granot Z, Geiss-Frielander R, Melamed-Brook N, Eimerl S, Timberg R, Weiss AM, Hales DB, Stocco DM, Orly J: Proteolysis of normal and mutated steroidogenic acute regulatory proteins in the mitochondria: The fate of unwanted proteins. Mol Endocrinol 2003; 17:2461–2476.
60 Arakane F, Sugawara T, Nishino H, Liu Z, Holt JA, Pain D, Stocco DM, Miller WL, Strauss JF 3rd: Steroidogenic acute regulatory protein (StAR) retains activity in the absence of its mitochondrial targeting sequence: Implications for the mechanism of StAR action. Proc Natl Acad Sci USA 1996;93:13731–13736.
61 Kallen CB, Billheimer JT, Summers SA, Staybrook SE, Lewis M, Strauss JF 3rd: Steroidogenic acute regulatory protein (StAR) is a sterol transfer protein. J Biol Chem 1998;273: 26285–26288.
62 Miller WL, Strauss JF 3rd: Molecular pathology and mechanism of action of the steroidogenic acute regulatory protein, StAR. J Steroid Biochem Mol Biol 1999;69:131–141.
63 Bose HS, Whittal RM, Baldwin MA, Miller WL: The active form of the steroidogenic acute regulatory protein, StAR, appears to be a molten globule. Proc Natl Acad Sci USA 1999;9: 7250–7255.
64 Christinsen K, Bose HS, Harris FM Miller WL, Bell JD: Binding of StAR to synthetic membranes suggests an active molten globule. J Biol Chem 2001;276:17044–17051.
65 Bose, HS, Lingappa VR, Miller WL: Rapid regulation of steroidogenesis by mitochondrial protein import. Nature 2002;417:87–91.
65a Yaworsky DC, Baker BY, Bose HS, Best KB, Jensen LB, Bell JD, Baldwin MA, Miller WL: pH-dependent interaction of the carboxyl-terminal helix of steroidogenic acute regulatory protein with synthetic membranes. J Biol Chem 2005;280:2045–2054.
66 Apter D, Pakarinen A, Hammond GL, Vihko R: Adrencortical function in puberty. Cortisol and DHEA in girls and boys. Acta Paediatr Scand 1979;68:599–604.
67 Sklar CA, Kaplan SL, Grumbach MM: Evidence for dissociation between adrenarche and gonadarche: Studies in patients with idiopathic precocious puberty, gonadal dysgenesis, isolated gonadotropin deficiency, and constitutionally delayed growth and adolescence. J Clin Endocrinol Metab 1980;51:548–556.
68 Orentreich N, Brind JL, Rizer RL, Vogelman JH: Age changes and sex differences in serum dehydroepiandrosterone sulfate concentrations throughout adulthood. J Clin Endocrinol Metab 1984;59:551–555.
69 Cutler GB, Glenn M, Bush M, Hodgen GD, Graham CE, Loriaux DL: Adrenarche: A survey of rodents, domestic animals and primates. Endocrinology 1978;103:2112–2118.

70 Arlt W, Martens JWM, Song M, Wang JT, Auchus RJ, Miller WL: Molecular evolution of adrenarche: Structural and functional analysis of P450c17 from four primate species. Endocrinology 2002;143:4665–4672.
71 Mellon SH, Shively JE, Miller WL: Human proopiomelanocortin (79–96), a proposed androgen stimulatory hormone, does not affect steroidogenesis in cultured human fetal adrenal cells. J Clin Endocrinol Metab 1991;72: 19–22.
72 Penhoat A, Sanchez P, Jaillard C, Langlois D, Begeot M, Saez JM: Human proopiomelanocortin (79–96), a proposed cortical androgen stimulating hormone, does not affect steroidogenesis in cultured human adult adrenal cells. J Clin Endocrinol Metab 1991;72:23–26.
73 Miller WL: The molecular basis of premature adrenarche: An hypothesis. Acta Paediatr 1999;88(suppl 433):60–66.
74 Zachmann M, Vollmin JA, Hamilton W, Prader A: Steroid 17,20 desmolase deficiency: A new cause of male pseudohermaphroditism. Clin Endocrinol 1972;1:369–385.
75 Goebelsmann U, Zachmann M, Davajan V, Israel R, Mestman JH, Mishell DR: Male pseudohermaphroditism consistent with 17,20-desmolase deficiency. Gynecol Invest 1976;7: 138–156.
76 Forest MG, Lecornu M, Peretti E: Familial male pseudohermaphroditism due to 17,20-desmolase deficiency. In vivo endocrine studies. J Clin Endocrinol Metab 1980;50:826–833.
77 Nakajin S, Shively JE, Yuan P, Hall PF: Microsomal cytochrome P-450 from neonatal pig testis: Two enzymatic activities (17α-hydroxylase and C17,20-lyase) associated with one protein. Biochemistry 1981;20:4037–4042.
78 Nakajin S, Hall PF: Microsomal cytochrome P-450 from neonatal pig testis. Purification and properties of a C_{21} steroid side-chain cleavage system (17α-hydroxylase-$C_{17,20}$ lyase). J Biol Chem 1981;256:3871–3876.
79 Nakajin S, Hall PF: Testicular microsomal cytochrome P450 for C_{21} steroid side-chain cleavage. J Biol Chem 1981;256:6134–6139.
80 Nakajin S, Shinoda M, Haniu M, Shively JE, Hall PF: C_{21} steroid side-chain cleavage enzyme from porcine adrenal microsomes. Purification and characterization of the 17α-hydroxylase/$C_{17,20}$ lyase cytochrome P-450. J Biol Chem 1984;259:3971–3976.
81 Kominami S, Shinzawa S, Takemori S: Purification and some properties of cytochrome P-450 for steroid 17α-hydroxylation and $C_{17,20}$ bond cleavage from guinea pig adrenal microsomes. Biochem Biophys Res Commun 1982; 109:916–921.
82 Suhara K, Fujimura Y, Shiroo M, Katagiri M: Multiple catalytic properties of the purified and reconstituted cytochrome P-450 (P-450sccII) system of pig testis microsomes. J Biol Chem 1984;259:8729–8736.
83 Zuber MX, Simpson ER, Waterman MR: Expression of bovine 17α-hydroxylase cytochrome P450 cDNA in non-steroidogenic (COS-1) cells. Science 1986;234:1258–1261.
84 Lin D, Harikrishna JA, Moore CCD, Jones KL, Miller WL: Missense mutation Ser[106]→ Pro causes 17α-hydroxylase deficiency. J Biol Chem 1991;266:15992–15998.
85 Chung B, Picado-Leonard J, Haniu M, Bienkowski M, Hall PF, Shivley JE, Miller WL: Cytochrome P450c17 (steroid 17α-hydroxylase/17,20 lyase): Cloning of human adrenal and testis cDNAs indicates the same gene is expressed in both tissues. Proc Natl Acad Sci USA 1987;84:407–411.
86 Picado-Leonard J, Miller WL: Cloning and sequence of the human gene encoding P450c17 (steroid 17α-hydroxylase/17,20 lyase): Similarity to the gene for P450c21. DNA 1987;6: 439–448.
87 Matteson KJ, Picado-Leonard J, Chung B, Mohandas TK, Miller WL: Assignment of the gene for adrenal P450c17 (17α-hydroxylase/17,20 lyase) to human chromosome 10. J Clin Endocrinol Metab 1986;63:789–791.
88 Fan YS, Sasi R, Lee C, Winter JSD, Waterman MR, Lin CC: Localization of the human CYP17 gene (cytochrome P450 17α) to 10q24.3 by fluorescence in situ hybridization and simultaneous chromosome banding. Genomics 1992;14:1110–1111.
89 Brentano ST, Picado-Leonard J, Mellon SH, Moore CCD, Miller WL: Tissue-specific, cyclic adenosine 3′,5′-monophosphate-induced, and phorbol ester-repressed transcription from the human P450c17 promoter in mouse cells. Mol Endocrinol 1990;4:1972–1979.
90 Lin CJ, Martens JWM, Miller WL: NF-1c, Sp 1, and Sp 3 are essential for transcription of the human gene for P450c17 (steroid 17α-hydroxylase/17,20 lyase) in human adrenal NCI-H295A cells. Mol Endocrinol 2001;15:1277–1293.
91 Flück CE, Miller WL: GATA-4 and GATA-6 modulate tissue specific transcription of the human gene for P450c17. Mol Endocrinol 2004;18:1144–1157.
92 Miller WL, Auchus RJ, Geller DH: The regulation of 17,20 lyase activity. Steroids 1997;62: 133–142.
93 Onoda M, Hall PF: Cytochrome b_5 stimulates purified testicular microsomal cytochrome P450 (C_{21} side-chain cleavage). Biochem Biophys Res Commun 1982;108:454–460.
94 Yanagibashi K, Hall PF: Role of electron transport in the regulation of the lyase activity of C-21 side-chain cleavage P450 from porcine adrenal and testicular microsomes. J Biol Chem 1986;261:8429–8433.
95 Kominami S, Ogawa N, Morimune R, Huang DY, Takemori S: The role of cytochrome b_5 in adrenal microsomal steroidogenesis. J Steroid Biochem Mol Biol 1992;42:57–64.
96 Tamburini PP, Gibson GG: Thermodynamic studies of the protein-protein interactions between cytochrome P-450 and cytochrome b_5. J Biol Chem 1983;258:13444–13452.
97 Guengerich FP, Martin M, Beaune P, Kremers P, Wolff T, Waxman D: Characterization of rat and human liver microsomal cytochrome P-450 forms involved in nifedipine oxidation, a prototype for genetic polymorphism in oxidative drug metabolism. J Biol Chem 1986;261:5051–5060.
98 Lin D, Black SM, Nagahama Y, Miller WL: Steroid 17α-hydroxylase and 17,20 lyase activities of P450c17: Contributions of serine[106] and P450 reductase. Endocrinology 1993;132:2498–2506.
99 Auchus RJ, Lee TC, Miller WL: Cytochrome b_5 augments the 17,20 lyase activity of human P450c17 without direct electron transfer. J Biol Chem 1998;273:3158–3165.
100 Yanase T, Sasano H, Yubisui T, Sakai Y, Takayanagi R, Nawata H: Immunohistochemical study of cytochrome b_5 in the human adrenal gland and in adrenocortical adenomas from patients with Cushing syndrome. Endocr J 1998;45:89–95.
101 Mapes S, Corbin C, Tarantal A, Conley A: The primate adrenal zona reticularis is defined by expression of cytochrome b_5, 17α-hydroxylase/17,20-lyase cytochrome P450 (P450c17) and NADPH-cytochrome P450 reductase (reductase) but not 3β-hydroxysteroid dehydrogenase/δ^{5-4} isomerase (3β-HSD). J Clin Endocrinol Metab1999;84:3382–3385.
102 Suzuki T, Sasano H, Takeyama J, Kaneko C, Freije WA, Carr BR, Rainey WE: Developmental changes in steroidogenic enzymes in human postnatal adrenal cortex: Immunohistochemical studies. Clin Endocrinol 2000;53: 739–747.
103 Yanase T, Waterman MR, Zachmann M, Winter JSD, Simpson ER, Kagimoto M: Molecular basis of apparent isolated 17,20 lyase deficiency: Compound heterozygous mutations in the C-terminal region (Arg(496)→Cys, Gln(461)→Stop) actually cause combined 17α-hydroxylase/17,20 lyase deficiency. Biochim Biophys Acta 1992;1139:275–279.
104 Geller DH, Auchus RJ, Mendonça BB, Miller WL: The genetic and functional basis of isolated 17,20 lyase deficiency. Nat Genet 1997; 17:201–205.
105 Geller DH, Auchus RJ, Miller WL: P450c17 mutations R347H and R358Q selectively disrupt 17,20-lyase activity by disrupting interactions with P450 oxidoreductase and cytochrome b_5. Mol Endocrinol 1999;13: 167–175.
106 Auchus RJ, Miller WL: Molecular modeling of human P450c17 (17α-hydroxylase/17,20-lyase): Insights into reaction mechanisms and effects of mutations. Mol Endocrinol 1999; 13:1169–1182.
107 Peterson RE, Imperato-McGinley J, Gautier T, Shackleton C: Male pseudohermaphroditism due to multiple defects in steroid biosynthetic microsomal mixed-function oxidases. A new variant of congenital adrenal hyperplasia. N Engl J Med 1985;313:1182–1191.

108 Miller WL: P450 oxidoreductase deficiency. A new disorder of steroidogenesis with multiple clinical manifestations. Trends Endocrinol Metab 2004 15:311–315.
109 Miller WL: Congenital adrenal hyperplasia (letter to editor). N Engl J Med 1986;314: 1321–1322.
110 Shen AL, O'Leary KA, Kasper CB: Association of multiple developmental defects and embryonic lethality with loss of microsomal NADPH-cytochrome P450 oxidoreductase. J Biol Chem 2002;277:6536–6541.
111 Otto DME, Henderson DJ, Carrie D, Davey M, Gunderson TE, Blomhoff R, Adams RH, Tickle C, Wolf CR: Identification of novel roles of they cytochrome P450 system in early embryogenesis: Effects on vasculogenesis and retinoic acid homeostatis. Mol Cell Biol 2003;23:6103–6116.
112 Reardon W, Smith A, Honour JW, Hindmarsh P, Das D, Rumsby G, Nelson I, Malcolm S, Ades L, Sillence D, Kumar D, DeLozier-Blanchet C, McKee S, Kelly T, McKeehan WL, Baraitser M, Winter RM: Evidence for digenic inheritance in some cases of Antley-Bixler syndrome? J Med Genet 2000;37:26–32.
113 Flück CE, Tajima T, Pandey AV, Arlt W, Okuhara K, Verge CF, Jabs EW, Mendonca BB, Fujieda K, Miller WL: Mutant P450 oxidoreductase causes disordered steroidogenesis with and without Antley-Bixler syndrome. Nature Genet 2004;36:228–230.
113a Huang N, Pandey AV, Agrawal V, Reardon W, Lapunzina PD, Mowat D, Jabs EW, Van Vliet G, Sack J, Flück CE, Miller WL: Diversity and function of mutations in P450 oxidoreductase in patients with Antley-Bixler syndrome and disordered steroidogenesis. Am J Hum Genet 2005;76:729–749.
114 Gazdar AF, Oie HK, Shackleton CH, Chen TR, Triche TJ, Myers CE, Chrousos GP, Brennan MF, Stein CA, LaRocca RV: Establishment and characterization of a human adrenocortical carcinoma cell line that expresses multiple pathways of steroid biosynthesis. Cancer Res 1990;50:5488–5496.
115 Staels B, Hum DW, Miller WL: Regulation of steroidogenesis in NCI-H295 cells: A cellular model of the human fetal adrenal. Mol Endocrinol 1993;7:423–433.
116 Zhang L, Rodriguez H, Ohno S, Miller WL: Serine phosphorylation of human P450c17 increases 17,20 lyase activity: Implications for adrenarche and for the polycystic ovary syndrome. Proc Natl Acad Sci USA 1995;92: 10619–10623.
117 Pandey AV, Mellon SH, Miller WL: Protein phosphatase 2A and phosphoprotein SET regulate androgen production by P450c17. J Biol Chem 2003;278:2837–2844.
118 Dunaif A: Hyperandrogenic anovulation (PCOS): A unique disorder of insulin action associated with an increased risk of non-insulin-dependent diabetes mellitus. Am J Med 1995;98(suppl 1A):33S–39S.
119 Lachelin GCL, Barnett M, Hopper BR, Brink G, Yen SSC: Adrenal function in normal women and women with the polycystic ovary syndrome. J Clin Endocrinol Metab 1979;49: 840–848.
120 Lucky AW, Rosenfield RL, McGuire J, Rudy S, Helke J: Adrenal androgen hyperresponsiveness to ACTH in women with acne and/or hirsutism: Adrenal enzyme defects and exaggerated adrenarche. J Clin Endocrinol Metab 1986;62:840–848.
121 Ibañez L, Potau N, Zampolli M, Prat N, Gussinye M, Saenger P, Vicens-Clavet E, Carrascosa E: Source localization of androgen excess in adolescent girls. J Clin Endocrinol Metab 1994;79:1778–1784.
122 Ehrmann DA, Barnes RB, Rosenfield RL: Polycystic ovary syndrome as a form of functional ovarian hyperandrogenism due to dysregulation of androgen secretion. Endocr Rev 1995;16:322–353.
123 Ibañez L, Potau N, Virdis R, Zampolli M, Terzi C, Gussinyé M, Carrascosa A, Vicens-Calvet E: Postpubertal outcome in girls diagnosed of premature pubarche during childhood: Increased frequency of functional ovarian hyperandrogenism. J Clin Endocrinol Metab 1993;76:1599–1603.
124 Oppenheimer E, Linder B, DiMartino-Nardi J: Decreased insulin sensitivity in prepubertal girls with premature adrenarche and acanthosis nigricans. J Clin Endocrinol Metab 1995;80:614–618.
125 Dunaif A, Segal K, Futterweit W, Dobrjansky A: Profound peripheral insulin resistance, independent of obesity, in the polycystic ovary syndrome. Diabetes 1989;38:1165–1174.
126 Ciaraldi TP, El-Roeiy A, Madar Z, Reichart D, Olefsky JM, Yen SCC: Cellular mechanisms of insulin resistance in polycystic ovarian syndrome. J Clin Endocrinol Metab 1992; 75:577–583.
127 Bollag G, Roth R, Beaudoin J, Mochley-Rosen D, Koshland D Jr: Protein kinase C directly phosphorylates the insulin receptor in vitro and reduces its protein-tyrosine kinase activity. Proc Natl Acad Sci USA 1986; 83:5822–5824.
128 Stadtmauer L, Rosen OM: Increasing the cAMP content of IM-9 cells alters the phosphorylation state and protein kinase activity of the insulin receptor. J Biol Chem 1986; 261:3402–3407.
129 Takayama S, White MF, Kahn CR: Phorbol ester-induced serine phosphorylation of the insulin receptor decreases its tyrosine kinase activity. J Biol Chem 1988;263:3440–3447.
130 Dunaif A, Xia J, Book C-B, Schenker E, Tang Z: Excessive insulin receptor serine phosphorylation in cultured fibroblasts and in skeletal muscle. J Clin Invest 1995;96:801–810.

Original Paper

Medical Principles and Practice

Med Princ Pract 2005;14(suppl 1):69–72
DOI: 10.1159/000086186

Received: November 17, 2004
Revised: March 19, 2005

Mutations Associated with Beta-Thalassemia intermedia in Kuwait

Adekunle Adekile[a] Mohammad Haider[a] Ferdane Kutlar[b]

[a]Department of Pediatrics, Faculty of Medicine, Kuwait University, Kuwait;
[b]Department of Medicine, Medical College of Georgia, Augusta, Ga., USA

Key Words
β-Thalassemia intermedia · Kuwait

Abstract
Objective: To identify the β-globin gene mutations associated with β-thalassemia (β-thal) intermedia in Kuwait. ***Subjects and Methods:*** Eighteen patients from 13 unrelated families, mean age 12.7 ± 8.1 years, range 4–31 years, were involved in the study. They did not require regular blood transfusion. Complete blood count and cation exchange high-performance liquid chromatography hemoglobin quantitation were carried out using standard techniques. β-Thal mutations were identified with a combination of PCR amplification, allele-specific oligonucleotide hybridization or direct DNA sequencing. The patients were also screened for the α_2-globin gene (–3.7 kb) deletion. ***Results:*** Of the 13 families, 4 were homozygous for the IVS-I-II (G→A) and 4 for the IVS-I-6 (T→C) mutations, while 1 each was a compound heterozygote for the following mutation combinations: CD 8 (–AA) and –101 (C→T); IVS-I-6 (T→C) and CD 19 (A→G); IVS-II-1 (G→A) and –28 (A→C); IVS-I-110 (G→A) and $\delta\beta^0$ deletion. Therefore, homozygosity for two typically mild mutations (IVS-II-1 and IVS-I-6) accounted for 61% of the genotypes in our patients. ***Conclusion:*** Our results indicate that screening should commence with these two common alleles in Kuwaiti patients presenting with β-thal syndrome. Early identification of intermedia patients will avoid the complications following an unnecessary hypertransfusion program.

Introduction

β-Thalassemia (β-thal) is a heterogeneous group of disorders associated with defects within the β-globin gene locus, leading to absent (β^0) or reduced (β^+) synthesis of the β-globin chain. Inheritance of only one of these β-thal alleles is usually asymptomatic (β-thal trait), but two alleles (homozygote or compound heterozygote) would produce thalassemia major (TM) or thalassemia intermedia (TI). While TM is associated with early onset of severe anemia, which rapidly becomes transfusion dependent, TI is milder and variable. The TI patients do not require regular transfusion but may have hepatosplenomegaly and skeletal abnormalities [1–3].

TI is also quite heterogeneous at the genotype level and several molecular mechanisms have been proposed. Typically, mutations in the promoter, splice site activation, consensus sequence and polyadenylation site of the β-globin gene are associated with mild or silent thalassemia. The –158 C→T substitution in the $^G\gamma$-globin gene promoter, linked to frameshift codon 6 (–A), codon 8 (–AA) and IVS-II-1 (G→A) is associated with increased hemoglobin (Hb) F synthesis and patients with these mutations commonly have TI. Coinheritance of α-thal trait is also associated with TI because it reduces globin chain imbalance [4–8].

There is usually a segregation of a small number of particular β-thal mutations in discrete populations or ethnic groups [9, 10]. However, in a situation where there is significant genetic admixture or interactions, the number of mutations may be quite high. It is therefore desirable

KARGER
Fax +41 61 306 12 34
E-Mail karger@karger.ch
www.karger.com

1011–7571/05/0147–0069$22.00/0

Accessible online at:
www.karger.com/mpp

Prof. Adekunle D. Adekile
Department of Pediatrics, Faculty of Medicine, Kuwait University
PO Box 24923
13110 Safat (Kuwait)
Tel. +965 531 9486, Fax +965 533 8940, E-Mail adekile@hsc.edu.kw

Table 1. Molecular and hematological characteristics of β-thal intermedia patients in Kuwait

Serial No.	Nationality	Family No.	Sex/age	Mutations	α-Globin genotype	Hb g/dl	MCV fl	MCH pg	A_2 %	Hb F %
1	Kuwaiti	1	M/17	IVS-II-1/IVS-II-1	–α/αα	9.9	85.3	23.8	1.3	98.7
2	Kuwaiti	2	M/16	IVS-II-1/IVS-II-1	–α/αα	9.2	72.6	20.9	3.9	96.1
3	Kuwaiti	2	M/13	IVS-II-1/IVS-II-1	–α/αα	11.0	88.8	25.8	2.7	95.9
4	Kuwaiti	3	F/31	IVS-II-1/IVS-II-1	αα/αα	9.0	74.5	21.2	2.0	98.0
5	Kuwaiti	4	F/4	IVS-II-1/IVS-II-1	αα/αα	10.0	78.3	25.4	1.0	99.0
6	Kuwaiti	5	F/13	IVS-I-6/IVS-I-6	αα/αα	9.0	69.7	18.3	5.7	23.5
7	Jordanian	6	F/4	IVS-I-6/IVS-I-6	αα/αα	9.1	60.7	16.7	3.7	31.0
8	Jordanian	7	F/18	IVS-I-6/IVS-I-6	αα/αα	8.7	64.5	19.0	5.2	24.1
9	Jordanian	8	M/15	IVS-I-6/IVS-I-6	αα/αα	8.2	63.2	17.2	6.4	12.3
10	Kuwaiti	9	M/10	–101/CD 8	αα/αα	10.9	59.1	17.4	5.7	25.6
11	Kuwaiti	9	M/20	–101/CD 8	αα/αα	9.9	60.3	17.9	5.3	22.1
12	Kuwaiti	9	M/29	–101/CD 8	αα/αα	11.3	58.7	17.6	7.4	23.7
13	Kuwaiti	10	F/8	CD 8/IVS-II-1	αα/αα	9.0	82.1	21.9	2.3	97.3
14	Kuwaiti	10	M/7	CD 8/IVS-II-1	–α/αα	10.4	77.4	22.1	1.5	98.5
15	Kuwaiti	11	M/6	IVS-I-6/CD 19	αα/αα	7.8	54.9	18.4	5.0	24.5
16	Kuwaiti	11	M/5	IVS-I-6/CD 19	ααα/αα	7.8	59.2	19.8	5.6	19.6
17	Iraqi	12	M/7	IVS-II-1/-28	–α/αα	10.2	71.6	22.0	2.7	54.0
18	Jordanian	13	F/5	IVS-I-110/δβ thal	αα/αα	8.2	67.3	24.5	1.1	92.7

MCV = Mean corpuscular volume; MCH = mean hemoglobin concentration; A_2 = hemoglobin A_2.

to establish a database of such mutations in every society in which β-thal is found. This not only facilitates prenatal and early childhood diagnosis, but also provides a solid basis for genetic counseling and the institution of appropriate management. Therefore, the aim of this study was to identify the β-globin gene mutations associated with β-thal intermedia in Kuwait.

Subjects and Methods

Patients referred to the Pediatric Hematology Clinic in the Mubarak and Amiri Hospitals, Kuwait, with the presumptive diagnosis of TI were eligible for inclusion into the study. The diagnosis of TI was based on the presence of chronic hypochromic, microcytic anemia with normal iron status. There were 18 patients (11 male and 7 female) from 13 families, 15 of whom were referred for investigation and the remaining 3 were siblings who were identified through family screening. Mean age was 12.1 + 8.1 years, range 4–31 years. Other inclusion criteria included a peripheral blood picture and Hb electrophoresis pattern consistent with β-thal syndrome and a baseline Hb level of ≥7.0 g/dl without the need for regular blood transfusion. The immediate family was screened with complete blood count and Hb electrophoresis and both parents had to have evidence of β-thal trait for the index case to be admitted into the study.

The 18 patients were fully characterized. Blood was obtained by venepuncture from each patient, complete blood count was done with an electronic cell counter (Coulter S) and Hb quantitation was done with cation exchange high-performance liquid chromatography [11]. DNA was isolated using the method of Poncz et al. [12]. The region of the β-globin gene was amplified with primers as previously described [13, 14]. They were screened for common β-thal mutations using allele-specific oligonucleotide hybridization. Those that were not readily identified had sequencing of the β-globin gene done using the dideoxy method of Sanger et al. [15], while one was solved by automated sequencing in Augusta, Ga., USA (F.K.) and one by gene mapping in London University Hospital, London, UK. All the patients were also screened for α-thal-2 –3.7-kb deletion using a PCR method [16]. Although full haplotyping was not done, all cases were screened for the $^{G}\gamma$-promoter Xmn I site (i.e. –158 C→T mutation) [13].

Results

The relevant characteristics of the β-thal intermedia patients including nationality, molecular and hematological parameters are listed in table 1. Thirteen were Kuwaitis, 4 Jordanians and 1 Iraqi.

The genotypes identified were homozygotes or compound heterozygotes for different combinations of alleles usually associated with silent or mild β-thal. The β-globin mutations are shown in table 1. The mild alleles, IVS-II-1 G→A and IVS-I-6 T→C accounted for 63.9% of all mutations. Other mutations included the silent –101 C→T, mild CD 19 A→G (Hb Malay), –28 (A→C), and CD 8 (–AA). There was a Sicilian $\delta\beta^0$ allele that was identified by gene mapping; the details of the methodology and specifics of the deletion will be reported separately.

Four families each had homozygosity for IVS-II-1 G→A and IVS-I-6 T→C mutation, i.e. 61.5% of the families studied. One family each had compound heterozygosity for the following combinations: CD 8 (–AA) and –101 (C→T); IVS-I-6 (T→C) and CD 19 (A→G); IVS-II-1 (G→A) and –28 (A→C); IVS-I-110 (G→A) and a Sicilian $\delta\beta^0$ deletion.

Of the 8 Kuwaiti families, 2 brothers (No. 15 and 16) with compound heterozygosity for IVS-I-6 and CD 19 (Hb Malay) had a Thai mother, while of the 4 Jordanian families, 3 (75%) were homozygous for the IVS-I-6 mutation.

The –158 T→C mutation in the $^{G}\gamma$-promoter was found in all the chromosomes carrying the following β-thal mutations: IVS-II-1 G→A, IVS-I-6 T→C and CD 8 (–AA). All others were negative. Only 5 patients had α-thal-2 deletion.

All the patients had a mild phenotype, none was ever on regular transfusion and only one, the IVS-I-110/δβ compound heterozygote, developed massive splenomegaly with hypersplenism. She required blood transfusion and splenectomy, after which she maintained a reasonable Hb level and was no longer transfused. Hb levels ranged from 7.8 to 11.3 g/dl (mean 9.4 ± 1.1) and Hb F from 12.3 to 98.6% (mean 57.9 ± 37.2)

Discussion

Although Kuwait is a small country in the northeast corner of the Arabian Peninsula, its population of about 2 million is very heterogeneous. It was founded by groups of families who migrated from the Najd region of Saudi Arabia in the 17th or 18th century [17]. However, some of the early settlers included people of Iranian and Iraqi ancestries. Failaka Island, located at the entrance to Kuwait bay, links present-day Kuwait to ancient civilizations [18, 19]. It was part of the Dilmun civilization, which flourished contemporaneously with the Indus Valley and Mesopotamian cultures, with which it had intimate contact, from 3000 to 1200 BC. More recently, it was used as a camp by Alexander the Great and his successors for about a century starting from 300 BC [20]. In more modern times, there has been intimate contact of Arabian Gulf indigenes with people from North and East Africa, the Mediterranean and East Asia. This varied background has had a significant impact on the genetic pool among Kuwaitis.

It is therefore not surprising that there is no discrete segregation of β-thal mutations in the Kuwaiti population. Indeed in a previous study, we found 12 different (mainly Mediterranean, Kurdish and Asian) mutations among our patients [13]. In a study from the United Arab Emirates, whose population is similar to Kuwait's, 35 mutations were reported [21]. This multiplicity of mutations makes it difficult to instate a simple model of screening to facilitate early or prenatal diagnosis. In lieu of this, it is advisable to document the mutations commonly associated with an intermedia phenotype. If the number of such is small, the initial screening can be limited to these. The family of a patient carrying such mutations can then be counseled as to the likely prognosis. Early diagnosis of TI is also important in order to avoid regular transfusions, as for TM, which may lead to significant morbidity from iron overload and other complications [6, 22].

Homozygosity involving the mild β^0 mutation, IVS-II-1 (G→A) and the β^+ mutation, IVS-I-6 (T→C) is associated with the intermedia phenotype in 8 (61.5%) of the 13 families in this study. These 2 along with the β^0 frameshift codon 8 (–AA), which was found (as a compound heterozygote) in 2 other families (No. 9 and 10), are all Mediterranean mutations known to be associated with a mild phenotype. They were found on chromosomes with the –158 C→T mutation on the $^{G}\gamma$-globin gene promoter. This is responsible for the increased production of Hb F, which in IVS-II-1 and CD 8 homozygotes is about 98% and in IVS-I-6 about 20% [5, 9, 10, 23].

There were 2 brothers (serial No. 15 and 16) who were compound heterozygotes for the IVS-I-6 (C→T) and the CD 19 (A→G) Malay mutations. These boys have a Kuwaiti father and a Thai mother. The CD 19 (A→G) is one of those mutations that activate an alternate splice site and is associated with an intermedia phenotype [24, 25]. –101 (C→T) mutation, found in 1 family, is typically associated with silent β-thal in that heterozygotes have normal hematological parameters (mean corpuscular volume, mean hemoglobin concentration), normal Hb A_2 and Hb F levels. The mutation was originally described from Turkish and Bulgarian families, but subsequently has been seen in Israelis and Italians [5, 26–28]. In the family in this study, the mutation was in compound heterozygosity with another mild mutation, CD 8 (–AA), which is why the patients had such a mild phenotype. The promoter mutation, –28 (A→C), found in compound heterozygosity with IVS-II-1 in 1 patient, is a β^+ mutant originally described in Kurdish Jews [29]. Interestingly, the patient is from Iraq, which has a sizable Kurdish population. In homozygotes, the mutation may be associated with TM, but when it is inherited along with a mild mutation, as in this patient, the intermedia phenotype prevails.

The only mutation encountered in this study, which is usually associated with a severe β^0 phenotype, is IVS-I-110 (G→A). However, the patient was a compound heterozygote with a δβ allele. The latter is associated with increased Hb F levels, which in this patient was 98%.

Conclusions

Since there are only a limited number of mutations associated with the TI phenotype among Kuwaitis, it is feasible to design a protocol to screen for these first in every patient presenting with an unclear picture. Strategies still have to be designed to quickly identify other mutations in all patients presenting with a severe thal picture to fully document all mutations encountered in this population.

Acknowledgements

The technical assistance of Dalal Himadri, Mark D'Souza and Alexandar Dimovski is appreciated. We thank our colleagues who referred their patients to us. This study was supported by Kuwait University Research award MK034.

References

1 Thein SL: β-Thalassemia. Baillieres Clin Haematol 1998;11:91–126.
2 Cao A, Galanello R, Rosatelli MC, Argiuolo F, de Virgilis S: Clinical experience of management of thalassemia: The Sardinian experience. Semin Hematol 1996;33:66–75.
3 Kazazian HH: The thalassemia syndromes: Molecular basis and prenatal diagnosis in 1990. Semin Hematol 1990;27:209–228.
4 Loukopoulos D: Thalassemias: Genotype and phenotypes. Ann Hematol 1991;62:145–150.
5 Kutlar A: The β-thalassemias: An overview; in Huisman THJ (ed): Current Views on Thalassemia. New York, Harwood Academic Publishers, 1992, pp 17–30.
6 Ratip S, Petrou M, Old JM, Wonke B, Porter JB, Modell B: Relationship between the severity of β-thalassaemia syndromes and the number of alleviating mutations. Eur J Haematol 1997;58:14–21.
7 Galanello R, Cao A: Relationship between genotype and phenotype: Thalassemia intermedia. Ann NY Acad Sci 1998;850:325–333.
8 Ho, PJ, Hall GW, Luo LY, Weatherall DJ, Thein SL: Beta-thalassaemia intermedia: Is it possible to predict phenotype from genotype? Br J Haematol 1998;100:70–78.
9 Hill AVS: Molecular epidemiology of the thalassaemias (including haemoglobin E). Baillieres Clin Haematol 1992;5:209–239.
10 Flint J, Harding RM, Boyce AJ, Clegg JB: The population genetics of haemoglobinopathies. Baillieres Clin Haematol 1993;6:215–261.
11 Bissé E, Wieland H: High-performance liquid chromatographic separation of human haemoglobins. Simultaneous quantitation of fetal and glycated hemoglobins. J Chromatogr 1988; 434:95–110.
12 Poncz M, Solowiejczyk D, Harpel B, Mory Y, Schwartz E, Surrey S: Construction of human gene libraries from small amounts of peripheral blood: Analysis of βS-like globin genes. Hemoglobin 1982;6:27–36.
13 Adekile AD, Gu L-H, Baysal E, Haider MZ, Al-Fuzae L, Aboobacker KC, Al-Rashied A, Huisman THJ: Molecular characterization of α-thalassaemia determinants, β-thalassaemia alleles and βS haplotypes among Kuwaiti Arabs. Acta Haematol 1994;92:176–181.
14 Gonzalez-Redondo JM, Stoming TA, Lanclos KD, Gu YC, Kutlar F, Kutlar F, Nakasuji T, Deng B, Han IS, Mckie VC, Huisman THJ: Clinical and genetic heterogeneity in Black patients with homozygous β-thalassemia from the southeastern United States. Blood 1988; 72:1007–1014.
15 Sanger F, Nicklen S, Coulson AR: DNA sequencing with chain termination inhibitors. Proc Natl Acad Sci USA 1977;74:5463–5469.
16 Baysal E, Huisman THJ: Detection of common deletional α-thalassemia-2 determinants by PCR. Am J Hematol 1994;46:208–213.
17 Abu-Hakima AM: The Modern History of Kuwait. London, Luzac, 1983, pp 1–10.
18 Adekile AD: Historical and anthropological correlates of βS haplotypes and α- and β-thalassemia alleles in the Arabian Peninsula. Hemoglobin 1997;21:281–296.
19 Nourel-Din SA: The Stamp Seals of Failaka: An Aesthetic Analysis. Kuwait, Department of Antiquity & Museums, Kuwait National Museum, 1993.
20 Abu-Hakima AM: History of Eastern Arabia 1750–1800: The Rise and Development of Bahrain, Kuwait and Wahhabi Saudi Arabia. London, Probsthain, 1988.
21 Al-Kalla S, Matthews AR: Molecular characterization of β-thalassemia in the United Arab Emirates. Hemoglobin 1993;17:355–362.
22 Ratip S, Skuse D, Porter J, Wonke B, Yardumian A, Modell B: Psychological and clinical aspects of thalassaemia intermedia and its implications for prenatal diagnosis. Arch Dis Child 1995;72:408–412.
23 Dedousis GVZ, Mandilara GD, Boussiu M, Loutradis A: HbF production in β thalassaemia heterozygotes for IVS-II-1 G→A mutation. Implication of the haplotype and the $^G\gamma$-158 C→T mutation on the HbF level. Am J Hematol 2000;64:151–155.
24 Yang KG, Kutlar F, George E, Wilson JB, Kutlar A, Stoming TA, Gonzalez-Redondo JM, Huisman THJ: Molecular characterization of beta-globin gene mutations in Malay patients with Hb E-beta-thalassaemia and thalassaemia major. Br J Haematol 1989;72:73–80.
25 Thein SL, Winichagoon P, Hesketh C, Best S, Fucharoen S, Wasi P, Weatherall DJ: The molecular basis of beta-thalassemia in Thailand: Application to prenatal diagnosis. Am J Hum Genet 1990;47:369–375.
26 Rund D, Oron-Karni V, Filon D, Goldfarb A, Rachmilewitz EA, Oppenheim A: Genetic analysis of β-thassemia intermedia in Israel: Diversity of mechanisms and unpredictability of phenotype. Am J Hematol 1997;54:16–22.
27 Maragoudaki E, Kanavakis E, Traeger-Synodinos J, Vrettou C, Tzdetis M, Metaxotou-Mavrommati A, Kattamis C: Molecular, haematological and clinical studies of the –101 C→T substitution of the β-globin gene promoter in 25 β-thalassaemia intermedia patients and 45 heterozygotes. Br J Haematol 1999;107:699–706.
28 Huisman THJ, Carver MFH, Baysal E: A Syllabus of Thalassemia Mutations. Augusta, The Sickle Cell Anemia Foundation, 1997, p 13.
29 Poncz M, Ballantine M, Solowiejczyk D, Barak I, Surrey S: Beta-thalassemia in a Kurdish Jew. Single base changes in the T-A-T-A box. J Biol Chem 1982;257:5994–5996.

Original Paper

Medical Principles and Practice

Med Princ Pract 2005;14(suppl 1):73–76
DOI: 10.1159/000086187

Received: November 17, 2004
Revised: February 28, 2005

Alpha-2-Globin Gene Polyadenylation (AATAAA→AATAAG) Mutation in Hemoglobin H Disease among Kuwaitis

M. Haider A. Adekile

Department of Pediatrics, Faculty of Medicine, Kuwait University, Kuwait

Key Words
Hemoglobin H disease · α_2-Globin gene polyadenylation · Kuwait

Abstract
Objectives: In the Arabian Gulf region, hemoglobin (Hb) H disease usually results from homozygosity or compound heterozygosity involving the α_2-globin gene polyadenylation (poly A) signal (AATAA$\underline{A}$→AATAA$\underline{G}$) mutation ($\alpha^T\alpha$). Here we document the clinical and hematological characteristics of children with Hb H disease being followed in Kuwait. ***Subjects and Methods:*** Twenty-four patients (0.5–12 years old, mean 4.7 ± 3.5 years) with persistent microcytic, hypochromic anemia (and normal iron status as well as normal Hb A_2 levels) were referred to the pediatric hematology clinic for further investigations. They were all screened for the α^+-thalassemia (α^+-thal; –3.7 kb) deletion using a standard PCR method. They were also screened for the α_2-globin gene $\alpha^T\alpha$ allele and the 5nt deletion ($-\alpha^{5nt}$) in the first intervening sequence, which are common α-thal alleles in this population. They were followed up for periods ranging from 2 to 8 years. ***Results:*** Of the 24 patients, 4 (16.7%) also had sickle cell trait (Hb-AS), while 7 (29.2%) were glucose-6-phosphate dehydrogenase deficient. Only 1 patient had significant hepatosplenomegaly and 1 developed gallstones. While none was on chronic transfusion therapy, 8 (33.3%) had been transfused at least once and, in 3 instances, this was secondary to parvovirus B19 +ve aplastic crisis. The α-globin genotype was successfully determined in almost all patients. The results showed that 17 (70.8%) patients were homozygous for the poly A mutation ($\alpha^T\alpha/\alpha^T\alpha$), 6 (25.0%) were compound heterozygotes for this and the α^+-thal (–3.7 kb) deletion ($-\alpha/\alpha^T\alpha$) and 1 (4.2%) was undetermined. There were no significant differences in the phenotypes of the 2 genotypes and their hematological features were identical. ***Conclusions:*** Hb H disease involving the poly A mutation is a mild thal intermedia phenotype among Kuwaitis. There are no serious complications and there is no need for regular blood transfusion.

Introduction

Normal individuals have two linked α-globin genes on chromosome 16, with the genotype αα/αα. The two most common determinants of α-thalassemia, α-thal-2 and α-thal-1, result from the loss of one (–α/) or both (––/) α-globin genes, respectively. The former is associated with α^+, while the latter is associated with α^0-thal trait. Compound heterozygosity of α^0-thal and α^+-thal alleles (––/–α) results in inactivation or loss of 3 α-globin genes,

KARGER
Fax +41 61 306 12 34
E-Mail karger@karger.ch
www.karger.com

1011–7571/05/0147–0073$22.00/0

Accessible online at:
www.karger.com/mpp

Prof. Adekunle Adekile
Department of Pediatrics, Faculty of Medicine, Kuwait University
PO Box 24923
13110 Safat (Kuwait)
Tel. +965 531 9486, Fax +965 533 8940, E-Mail adekile@hsc.kuniv.edu.kw

with significant α/β-globin chain imbalance causing the clinical syndrome called hemoglobin (Hb) H disease. Less common α-thal determinants are nondeletional lesions in which a mutation on one or the other α-globin gene ($\alpha^T\alpha/$) reduces the synthetic production of the affected gene. Homozygotes ($\alpha^T\alpha/\alpha^T\alpha$) or compound heterozygotes ($-\alpha/\alpha^T\alpha$, $--/\alpha^T\alpha$) are also associated with the Hb H phenotype of varying severity [1–4].

The clinical phenotype of Hb H disease is quite heterogeneous, but is usually one of thal intermedia with mild to moderate microcytic, hypochromic anemia, usually not requiring regular blood transfusion. The excess β-globin chains form unstable tetramers (Hb H), which precipitate to form Hb H inclusion bodies in peripheral red blood cells. An H band may also be demonstrable on electrophoresis or high-performance liquid chromatography. In the newborn period, γ-tetramers (Hb Bart's) are seen. There may be varying degrees of hepatosplenomegaly and skeletal deformities [3–6].

α-Thal trait is quite prevalent in the Arabian Peninsula with the frequency of α-thal-2 allele being as high as 40–60% in some areas [7–9]. While the α^0-thal alleles are uncommon, the nondeletional, α_2-globin polyadenylation (poly A; AATAA<u>A</u>→AATAA<u>G</u>) allele has been described in Eastern Saudi Arabia, United Arab Emirates and Kuwait [3, 4, 10–12]. There have been mainly anecdotal reports of Hb H disease among Gulf Arabs [8–10], and the present communication is an account of cases that are being followed in the pediatric hematology clinic in Kuwait. It documents their clinical presentation, laboratory findings and molecular characterization. In addition, it looks at the influence (if any) of coexistent sickle cell trait and glucose-6-phosphate dehydrogenase (G6PD) deficiency on the phenotype.

Subjects and Methods

A total of 24 patients were studied, 14 males and 10 females aged between 6 months and 12 years with a mean age of 4.7 ± 3.6 years. Eleven (45.8%) had weight and height of ≥50th centile. Their Hb ranged from 7.4 to 10.6 g/dl (9.0 ± 0.8), mean cell volume from 48.1 to 66.6 fl (56.7 ± 4.0) and mean cell hemoglobin from 15.7 to 21.2 pg (18.4 ± 1.4). All patients had inclusion bodies in their peripheral blood smears.

The patients were being followed in the pediatric hematology clinics of the Amiri and Mubarak Hospitals in Kuwait. Both are part of the teaching hospitals' complex of the Kuwait University Faculty of Medicine. The diagnosis of Hb H disease was based on the finding of microcytic, hypochromic anemia in a child who did not have iron deficiency and whose Hb electrophoresis was not consistent with the β-thal syndrome. In addition, they all showed Hb H inclusion bodies in more than 80% of the peripheral red blood cells following cresyl blue staining. Family screening was carried out and in all cases both parents had varying degrees of microcytic and hypochromic anemia.

Blood was drawn from all the patients by venepuncture; complete blood count was done with the Coulter electronic cell counter. A qualitative test for G6PD deficiency was used for screening the patients [13]. DNA was extracted by the method of Poncz et al. [14]. All the patients were screened for the α-thal-2 ($-\alpha^{-3.7}$ kb) deletion with a PCR method [15]. The α_2-globin gene region was again amplified and dot blotting was done followed by allele-specific hybridization with probes specific for the poly A mutation [4, 11], AATAA<u>A</u>→AATAA<u>G</u> ($\alpha^T\alpha/$) and 5nt deletion ($-\alpha^{-5nt}$) in the 5′ splice junction of the first intervening sequence [16]. These are the two nondeletional α-thal alleles prevalent in this region [10–12].

All the patients had detailed physical examination and were followed regularly in the clinic. The patients received daily folic acid supplement; otherwise, no other routine treatment was given. Their charts were reviewed for this report and details of any hospitalization were obtained. Particular attention was paid to frequency of blood transfusion and any complications, e.g. aplastic crisis.

Results

Of the 24 patients studied, 7 (29.2%) had G6PD deficiency, while 3 (12.5%) had Hb AS (heterozygous normal /sickle-Hb). Eight (33.3%) had been transfused at least once and none was on regular transfusion. Four (57.1%) of the G6PD-deficient patients had been transfused; the difference in frequency was not significant ($\chi^2 = 1.0$, $p > 0.05$). There were also no other significant differences in the hematological parameters among those with G6PD deficiency or Hb AS; this may be due to the small sample size.

Seventeen (70.8%) were homozygous for the poly A mutation ($\alpha^T\alpha/\alpha^T\alpha$) and 6 (25.0%) were compound heterozygous for the α-thal-2 deletion and the poly A mutation ($-\alpha^{3.7}/\alpha^T\alpha$), while 1 (4.2%) was undetermined. There were no significant differences in the hematological parameters among the homozygotes and the compound heterozygotes.

The patients did not show any common complication, but 1 (4.2%), with genotype $\alpha^T\alpha/\alpha^T\alpha$, who was also G6PD deficient, had persistent hepatosplenomegaly, first noticed when he was about 6 months old and persisting at age 9 years. He was also transfused on 3 occasions in the first 2 years of life. Three (12.5%) patients developed aplastic crisis secondary to parvovirus B19 infection. Four (16.7%) patients had reactive thrombocytosis with platelet count ranging from 600 to >1,000 × 10^9/l lasting for varying lengths of time. None had evidence of concurrent iron deficiency. Only 1 (4.2%) patient had gallstones and required cholecystectomy.

Discussion

Hb H disease is indeed common in Kuwaitis with ancestral origin in Eastern Saudi Arabia [17]. The patients are usually referred to the pediatric hematology clinic because of significant hypochromic, microcytic anemia not responding to iron supplements and with normal Hb A_2 levels, ruling out β-thal trait. Almost all cases involve the α_2-globin gene AATAA$\underline{A}$→AATAA$\underline{G}$ mutation either as homozygotes or compound heterozygotes with the $-\alpha^{3.7\ kb}$ α-thal deletion.

The phenotype of Hb H disease among our patients is uniformly mild, with none of them requiring regular transfusions. Patients with chronic hemolytic anemia are prone to aplastic crisis following documented parvovirus B19 infection [18–20]. This was encountered in 3 of our patients; it was self-limiting in all cases, although they all required at least one blood transfusion. Four of the patients in this study showed evidence of transient reactive thrombocytosis. This appears to be nonspecific and accompanies the anemia of the disease.

Concomitant G6PD deficiency may be responsible for frequent hemolytic crisis in sickle cell disease. Its effect, if any, on the phenotype in Hb H disease has not been widely documented. Sanna et al. [21] reported higher G6PD enzyme levels in patients with Hb H disease and Öner et al. [22] did not see any effect of the enzyme deficiency on the hematological parameters of their patient. Similarly, we could not demonstrate any significant effects of G6PD deficiency on our patients and the frequency of blood transfusion was not significantly higher.

Matthay et al. [23] have suggested that sickle cell trait may modify Hb H disease because the relatively low Hb A level reduces the availability of β-chains to form Hb H. On the other hand, α-thal trait is known to ameliorate the course of sickle cell disease, because intracellular concentration of Hb S is low in patients with coexistent α-thal trait [24, 25]. It was therefore interesting that 3 of the Hb H patients in the present study had sickle cell trait. Two of them had Hb H inclusions in more than 80% of their peripheral erythrocytes and 1 had in about 40%. However, the baseline Hb level and other hematological parameters were not significantly different. We therefore cannot make a case for a modifying role for sickle cell trait in our Hb H patients.

No specific treatment is offered to our Hb H disease patients. Full family screening is carried out on all suspected cases to document complete cell count, iron status, Hb electrophoresis and α-globin genotype. Genetic counseling is carried out and the affected patients are started on regular folic acid supplementation. Patients who maintain their Hb at 10 g/dl and above are not followed in the clinic.

Conclusions

Hb H disease involving the poly A mutation is a mild thal phenotype among Kuwait patients. There are no serious complications and they do not require regular blood transfusion.

Acknowledgements

The technical assistance of Mark D'Souza, Himadri Dalal and Jalaja Sukumaran is appreciated. This work was supported by Kuwait Research Administration grant MK034.

References

1 Higgs DR, Vickers MA, Wilkie AOM, Pretorius IM, Jarman AP, Weatherall DJ: A review of the molecular genetics of the human α-globin gene cluster. Blood 1989;73:1081–1104.
2 Kazazian HH: The thalassemia syndromes: Molecular basis and prenatal diagnosis in 1990. Semin Hematol 1990;27:209–228.
3 Pressley L, Higgs DR, Clegg JB, Perrine RP, Pembrey ME, Weatherall DJ: A new genetic basis for hemoglobin-H disease. N Engl J Med 1980;303:1383–1388.
4 Thein SL, Wallace RB, Pressley L, Clegg DJ, Weatherall DJ, Higgs DR: The polyadenylation site mutation in the α-globin gene cluster. Blood 1988;71:313–319.
5 Lui TC, Chiou SS, Lin SF, Chen TP, Tseng WP, Chen PH, Chang JG: Molecular basis and hematological characterization of Hb H disease in Southeast Asia. Am J Hematol 1994; 45:293–297.
6 Weatherall DJ, Clegg JB: The Thalassemia Syndromes, ed 3. Oxford, Blackwell Publishing, 1981.
7 El-Hazmi MAF: Haemoglobin disorders: A pattern of thalassaemia and haemoglobinopathies in Arabia. Acta Haematol 1982;68:43–51.
8 White JM, Byrne M, Richards R, Buchanan T, Katsoulis E, Weerasingh K: Red cell genetic abnormalities in Peninsular Arabs: Sickle haemoglobin, G6PD deficiency, and alpha and beta thalassaemia. J Med Genet 1986;23:245–251.
9 El-Hazmi MAF: Hemoglobinopathies, thalassemias and enzymopathies in Saudi Arabia. Saudi Med J 1992;13:488–499.
10 El-Kalla S, Baysal E: Alpha thalassemia in the United Arab Emirates. Acta Haematol 1998; 100:49–53.
11 Adekile AD, Gu, L-H, Baysal E, Haider MZ, Al-Fuzae L, Aboobacker KC, Al-Rashied A, Huisman THJ: Molecular characterization of α-thalassaemia determinants, β-thalassaemia alleles and βS haplotypes among Kuwaiti Arabs. Acta Haematol 1994;92:176–181.
12 Adekile AD, Haider MZ: Morbidity, βS haplotype and α-globin gene patterns among sickle cell anemia patients in Kuwait. Acta Haematol 1996;96:150–154.

13 Beutler E, Blume KG, Kaplan JC: International Committee for Standardization in Haematology: Recommended screening test for glucose-6-phosphate dehydrogenase (G6PD) deficiency. Br J Haematol 1979;43:469–477.
14 Poncz M, Solowiejczyk D, Harpel B, Mory Y, Schwartz E, Surrey S: Construction of human gene libraries from small amounts of peripheral blood: Analysis of βS-like globin genes. Hemoglobin 1982;6:27–36.
15 Baysal E, Huisman THJ: Detection of common deletional α-thalassemia-2 determinants by PCR. Am J Hematol 1994;46:208–213.
16 Orkin SH, Goff SC, Hechtman RL: Mutation in an intervening sequence splice junction in man. Proc Natl Acad Sci USA 1981;78:5041–5045.
17 Adekile AD: Historical and anthropological correlates of βS haplotypes and α- and β-thalassemia alleles in the Arabian Peninsula. Hemoglobin 1997;21:281–296.
18 Hayakawa A, Sano K, Hasegawa D, Kosaka Y, Nakamura H, Ideguchi H, Harano T: Hereditary hemoglobin H disease in Japanese siblings diagnosed by human parvovirus B19 infection. Rinsho Ketsueki 1999;40:536–541.
19 Kellermayer R, Faden H, Grossi M: Clinical presentation of parvovirus B19 infection in children with aplastic crisis. Pediatr Infect Dis J 2003;22:1100–1101.
20 Badr MA: Human parvovirus B19 infection among patients with chronic blood disorders. Saudi Med J 2002;23:295–297.
21 Sanna G, Frau F, Melis MA, Galanello R, De Virgiliis S, Cao A: Interaction between the glucose-6-phosphate dehydrogenase deficiency and thalassaemia genes at phenotype level. Br J Haematol 1980;44:555–561.
22 Öner C, Öner R, Birben E, Balkan H, Gümruk F, Gürgey A, Altay Ç: Hb H disease with homozygosity for red cell G6PD deficiency in a Turkish female. Hemoglobin 1998;22:157–160.
23 Matthay KK, Mentzer WC, Dozy AM, Kan YW and Bainton DF: Modification of hemoglobin H disease by sickle cell trait. J Clin Invest 1979;64:1024–1032.
24 Adams RJ, Kutlar A, McKie V, Carl E, Nichols FT, Liu JC, Mckie K, Clary A: Alpha thalassemia and stroke in sickle cell anemia. Am J Hematol 1994;45:279–282.
25 Steinberg MH: Modulation of the phenotypic diversity of sickle cell anemia. Hemoglobin 1996;20:1–19.

Original Paper

Medical Principles and Practice

Med Princ Pract 2005;14(suppl 1):77–83
DOI: 10.1159/000086188

Received: October 19, 2004
Revised: March 1, 2005

Phenotypic and Molecular Characterization of *Candida dubliniensis* Isolates from Clinical Specimens in Kuwait

Suhail Ahmad Eiman Mokaddas Noura Al-Sweih Zia U. Khan

Department of Microbiology, Faculty of Medicine, Kuwait University, Kuwait

Key Words
Candida dubliniensis · *Candida albicans* · Isolation · Molecular characterization · rDNA sequencing

Abstract
Objective: This study was carried out to characterize *Candida dubliniensis* using phenotypic and molecular methods and to determine the occurrence of *C. dubliniensis* in clinical specimens in Kuwait. ***Materials and Methods:*** A total of 880 clinical specimens for isolation of fungi were processed according to standard procedures. Of these, 390 germ-tube-positive clinical isolates of *Candida* species were examined for rough colonies with hyphal fringes and chlamydospore production on simplified sunflower seed agar for their presumptive phenotypic identification as *C. dubliniensis.* The identification of *C. dubliniensis* isolates was further confirmed by the Vitek 2 yeast identification system, semi-nested (sn) PCR amplification of high-copy rDNA and direct DNA sequencing of the internally transcribed spacer 2 (ITS2) region. ***Results:*** Of the 390 isolates of *Candida* species investigated, 12 were identified as *C. dubliniensis,* giving an overall occurrence of 3%. All the *C. dubliniensis* isolates formed rough colonies with hyphal fringes and abundant chlamydospores on sunflower seed agar, did not assimilate trehalose, lactate and α-methyl-*D*-glucoside, and were isolated from human immunodeficiency virus (HIV)-negative patients. Four *C. dubliniensis* isolates utilized *D*-xylose. The species-specific primer derived from the ITS2 sequence of *C. dubliniensis* and used together with the panfungal reverse primer in the reamplification step of the snPCR specifically amplified rDNA from reference and clinical *C. dubliniensis* isolates and not from *C. albicans* or other *Candida* species. The identity of two representative isolates was confirmed by DNA sequencing of the ITS2 region. ***Conclusions:*** The identity of 12 *C. dubliniensis* isolates was first established by phenotypic characteristics and then by snPCR using species-specific primers derived from ITS2 sequences. The recovery of *C. dubliniensis* from HIV-negative patients from Kuwait reinforces the existing view that this novel yeast species has a worldwide distribution and its occurrence is not restricted to any particular immunocompromised population.

Introduction

Candida dubliniensis is a newly recognized species that is genotypically distinct from *C. albicans* but shares several phenotypic characteristics with the latter species [1]. It occurs as a minor constituent of the human oral microflora [2]. In the past, *C. dubliniensis* isolates have been misidentified as *C. albicans*, since both species are germ tube positive and produce chlamydospores on cornmeal Tween 80 agar [1–3]. Although initially *C. dubliniensis* was believed to be associated with oropharyngeal candidiasis in human immunodeficiency virus (HIV)-infected patients [4, 5], recent reports suggest that this species has a much wider epidemiologic significance [2]. The species

KARGER

Fax +41 61 306 12 34
E-Mail karger@karger.ch
www.karger.com

1011–7571/05/0147–0077$22.00/0

Accessible online at:
www.karger.com/mpp

Prof. Zia U. Khan
Department of Microbiology, Faculty of Medicine, Kuwait University
PO Box 24923, 13110 Safat (Kuwait)
Tel. +965 531 2300, ext. 6505, Fax +965 531 8454, +965 533 2719
E-Mail ziauddin@hsc.edu.kw

has been isolated from a variety of clinical specimens and anatomic sites of HIV-negative as well as HIV-positive individuals with and without symptoms of oral candidiasis [4–8].

Given the growing importance of *C. dubliniensis* as an opportunistic pathogen [7–9], development of rapid methods to discriminate *C. dubliniensis* from *C. albicans* is highly desirable. Recently, several seed-based and non-seed-based media have been described to discriminate *C. dubliniensis* from *C. albicans* on the basis of colony characteristics and chlamydospores production [10–15]. Rapid identification of clinical *C. dubliniensis* isolates has also been described by the use of specific primers in PCR [16, 17].

In this study, we report isolation and identification of 12 isolates of *C. dubliniensis* from non-HIV-infected patients from Kuwait during a 1-year period. The identification of the *C. dubliniensis* isolates was performed by phenotypic characteristics and by a semi-nested PCR (snPCR) using panfungal and *C. dubliniensis*-specific primers as described in the previous studies [18–20].

Materials and Methods

Isolation and Phenotypic Characterization

A total of 880 clinical specimens (vaginal swab: 410; urine: 123; bronchoalveolar lavage: 85; sputum: 81; cerebrospinal fluid: 46; blood culture: 40; tracheal secretion: 36; peritoneal and gastric fluid: 21; throat swab: 14; wound swab: 7; catheter tip: 7; tissue biopsies: 6, and bile fluid: 4) for the isolation of fungi received at the Mycology Laboratory during January to December 2003 were processed according to standard procedures as previously described [21]. All the yeast isolates (n = 452) were streaked on Sabouraud glucose agar (glucose 15 g, peptone 10 g and agar 15 g in 1 liter distilled water, pH 6.8) supplemented with chloramphenicol (50 mg/l) and checked for purity. The inoculum from the isolated yeast colonies was tested for germ tube formation in serum at 37°C. The germ tube-positive isolates (390) were examined for colonial and microscopic characteristics on simplified sunflower seed agar as previously described [10]. Isolates producing rough colonies with hyphal fringes and chlamydospores were provisionally identified as *C. dubliniensis.* They were further identified by their inability to assimilate trehalose and α-methyl-*D*-glucoside in the Vitek 2 yeast identification system (BioMerieux, Marey l'Etoile, France). Isolates were also tested for tolerance to 6.5% sodium chloride to distinguish *C. dubliniensis* from *C. albicans* [22].

Molecular Characterization

C. dubliniensis (CD 36), *C. albicans* (ATCC 76615), *C. parapsilosis* (ATCC 10233), *C. tropicalis* (ATCC 750), *C. glabrata* (ATCC 15545), *C. krusei* (clinical isolate) and *C. lusitaniae* (clinical isolate) were used as reference *Candida* species for snPCR amplification of rDNA. In addition, DNA extracted from *Aspergillus fumigatus* (ATCC 13073), *A. terreus* (CBS 118.27), *Fusarium solani* (ATCC 36031), *Trichosporon asahii* CBS 2530, *Cryptococcus neoformans* (ATCC 90112), *Toxoplasma gondii* (clinical isolate), human cells, *Mycobacterium tuberculosis* $H_{37}Rv$ and *Escherichia coli* BL-21 was also included in the study.

snPCR Amplification Method

The genomic DNA was extracted from the grown cultures of the isolates as previously described [18, 19]. The sequences of the universal forward and reverse fungal primers, capable of amplifying the 3′ end of 5.8S and 5′ end of 28S rDNA including the internally transcribed spacer 2 (ITS2) region and the species-specific oligonucleotide primer, derived from the ITS2 region of *C. dubliniensis,* have already been described [18–20]. Amplification with universal fungal primers and species-specific primer together with universal fungal reverse primer in the first round and snPCR from *C. dubliniensis* CD 36 should yield DNA fragments of 350 bp and 105 bp, respectively [20]. Amplification of target DNA was carried out in thin-walled 0.2-ml PCR tubes in a total volume of 50 μl containing 1× AmpliTaq PCR buffer I, 1 U AmpliTaq DNA polymerase, 10 pmol each of CTSF and CTSR primers, 1 μl of DNA and 0.1 m*M* of each dNTP. After amplification in the first step, 1 μl of the product was further amplified using the initial reverse primer (CTSR) and *C. dubliniensis*-specific forward primer. For snPCR, the reaction mixture consisted of 1× AmpliTaq PCR buffer I, 1 U AmpliTaq DNA polymerase, 5 pmol of CTSR together with 5 pmol of CDDET; 1 μl of the first PCR product and 0.1 m*M* of each dNTP. PCR cycling (total 30 cycles) was carried out in a Perkin-Elmer cycler (GeneAmp PCR system 2400) under the following conditions: denaturation at 94°C for 1 min, annealing at 60°C for 30 s and extension at 72°C for 1 min. An initial denaturation step at 95°C for 3 min and a final extension step at 72°C for 10 min were also included. To detect amplified DNA fragments, agarose gel electrophoresis was performed as previously described [18]. The gels were exposed to UV light and photographed. Appropriate negative controls were included in each test run, including controls omitting the DNA template during PCR assays.

The DNA sequence of the ITS2 region of two representative clinical *C. dubliniensis* isolates from Kuwait (MF 1042 and MF 122) was determined by amplification of the ITS2 region with universal fungal primers. The sequencing reactions with amplified DNA as template and forward or reverse panfungal primer were performed as described elsewhere [23]. Briefly, the column-purified PCR-amplified fragment obtained with the panfungal primer pair was used as template in sequencing reactions using the cycle DNA sequencing kit (DTCS CEQ2000, Beckman Coulter). The reaction mixtures in a final volume of 20 μl contained eluted DNA, 5 pmol of CTSF or CTSR primer and 8.0 μl of the premix reaction components supplied in the kit. The cycling parameters for the sequencing reactions included an initial denaturation at 95°C for 5 min followed by 30 cycles of 1 min at 94°C, 30 s at 55°C and 2 min at 60°C. The reaction products were precipitated and loaded on the sequencer as directed by the manufacturer of the automated DNA sequencer (Beckman Coulter Model CEQ8000).

Antifungal Susceptibility Testing

Antifungal susceptibility of the 12 *C. dubliniensis* isolates was determined against amphotericin B, fluconazole and voriconazole by E-test (AB BIODISK, Solna, Sweden). Each *C. dubliniensis* isolate was freshly subcultured and checked for purity. Five isolated colonies were uniformly suspended in sterile saline solution and

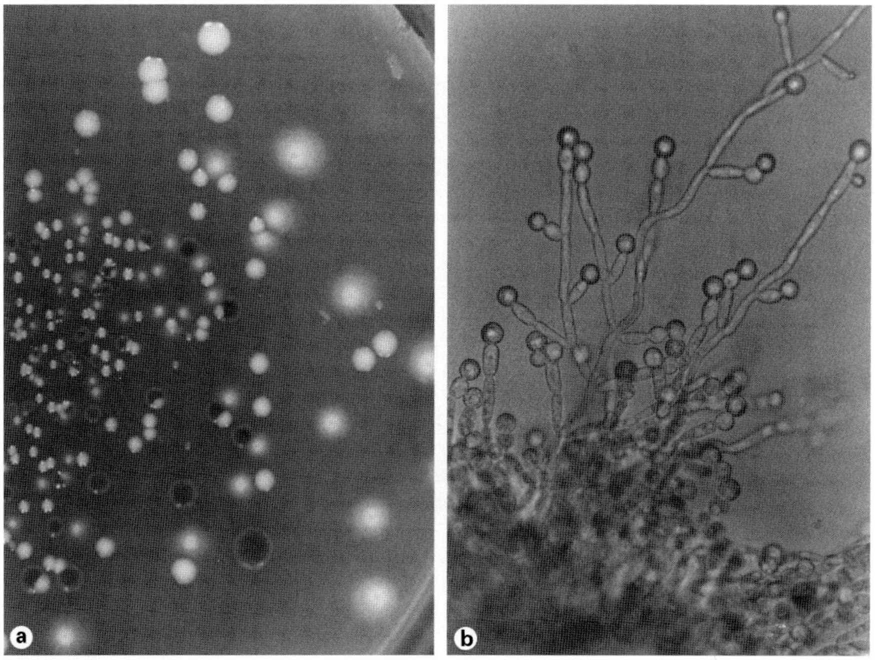

Fig. 1. **a** Simplified sunflower seed agar showing white rough colonies with hyphal fringes of *C. dubliniensis,* white smooth colonies of *C. albicans* and dark colonies of *C. neoformans* cultured from experimentally seeded bronchoalveolar lavage sample. **b** Slide culture on simplified sunflower seed agar showing chlamydospores of *C. dubliniensis* (magnification ×400).

turbidity was adjusted to 0.5 on the McFarland standard scale. This inoculum was then streaked onto the Sabouraud dextrose agar Petri plates (150 mm diameter) and allowed to dry at room temperature for 10–15 min before the application of the E-test strips. The RPMI-1640 agar supplemented with 2% glucose and buffered with MOPS, 0.165 *M*, pH 7.0, was used for antifungal susceptibility testing (National Committee for Clinical Laboratory Standards, NCCLS, M-27A) [24]. The plates were incubated in an incubator at 35°C for 1–2 days and minimal inhibitory concentrations (MICs) were read after 24–48 h of incubation. The point where inhibition ellipses intercepted the scale on the antifungal strip was taken as MIC for each test isolate. Complete inhibition (100%) of growth for amphotericin B and marked decrease in growth intensity for fluconazole and voriconazole were observed [25, 26].

Results

Of the 390 germ tube-positive isolates of *Candida* species, 12 were identified as *C. dubliniensis* based on colonial and microscopic characteristics on simplified sunflower seed agar medium. All the 12 *C. dubliniensis* isolates formed rough colonies with hyphal fringes and abundant chlamydospores on this medium after 48 h of incubation at 30°C (fig. 1a, b), while none of the remaining 378 isolates exhibited these characteristics. The results of phenotypic and genotypic characterization and the antifungal susceptibility profile are presented in tables 1, 2 and figures 1–3. Among the 12 isolates of *C. dubliniensis*, 9 were recovered from the respiratory tract and 1 each from urine, vaginal swab and catheter tip. All the 12 *C. dubliniensis* isolates originated from HIV-negative patients and none of these patients had systemic *Candida* infection. None of the 12 *C. dubliniensis* isolates assimilated α-methyl-*D*-glucoside, lactate and trehalose after 24 h of incubation in the Vitek 2 yeast identification system. Four *C. dubliniensis* isolates assimilated *D*-xylose. The growth of *C. dubliniensis* isolates was inhibited in the presence of 6.5% sodium chloride (table 1).

The results of the E-test performed on the 12 *C. dubliniensis* isolates are presented in table 2. The E-test MIC ranges for the three antifungal agents were as follows: amphotericin B 0.002–0.038 μg/ml, fluconazole 0.094–0.75 μg/ml, and voriconazole 0.003–0.012 μg/ml. None of *C. dubliniensis* isolate was resistant to any of the three antifungal agents tested.

The PCR amplification of rDNA from the 7 *Candida* species, i.e., *C. albicans, C. dubliniensis, C. parapsilosis, C. tropicalis, C. glabrata, C. krusei* and *C. lusitaniae,* using universal fungal primers (CTSF and CTSR) resulted in amplification of DNA fragments of expected sizes (fig. 2a). Similar results were obtained when the genomic DNA prepared from *A. fumigatus, F. solani, T. asahii* and *C. neoformans* was used as template (data not shown). Reamplification of the product of the first PCR with CTSR and CDDET resulted in specific amplification of a single DNA fragment of the expected size from the reference strain of *C. dubliniensis* only and not from the 6

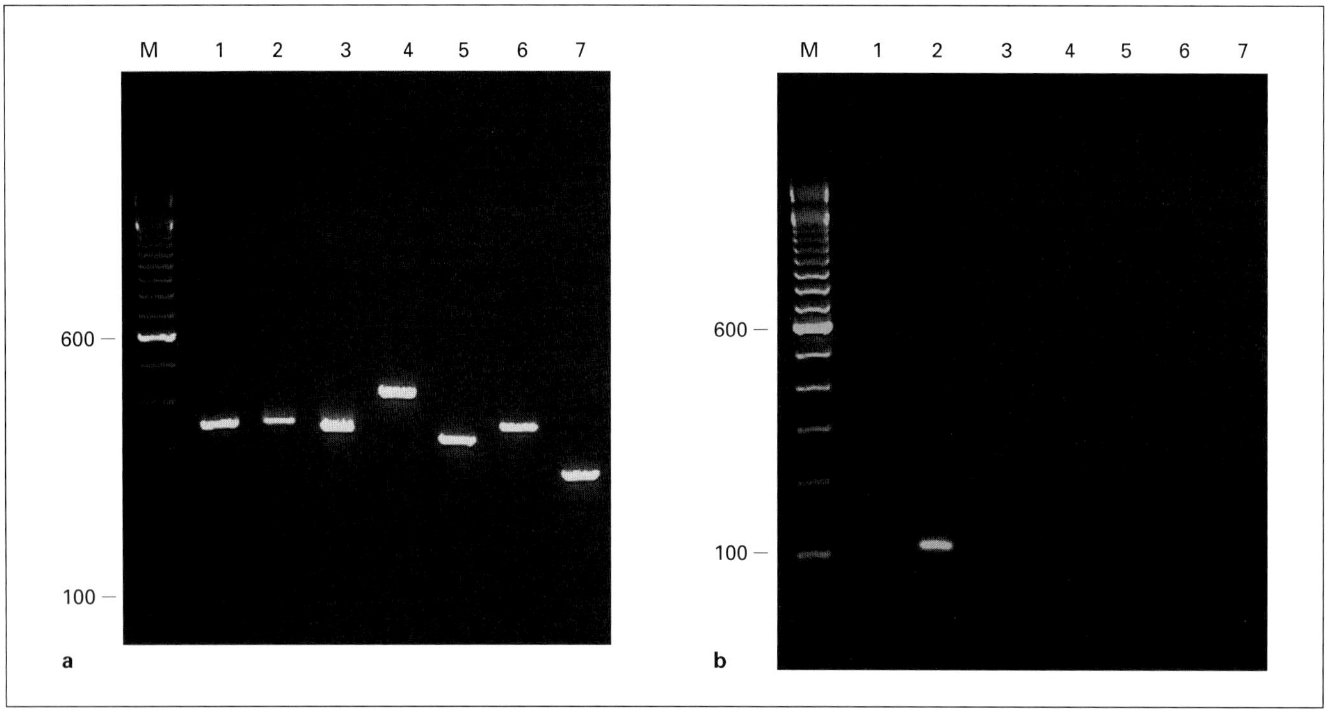

2

Fig. 2. Agarose gel electrophoresis of the product of the first round of PCR amplification using universal fungal primers (CTSF and CTSR) (**a**) and snPCR amplification using CTSR and CDDET primers (**b**) of genomic DNA from *C. albicans* (lane 1), *C. dubliniensis* (lane 2), *C. tropicalis* (lane 3), *C. glabrata* (lane 4), *C. parapsilosis* (lane 5), *C. krusei* (lane 6) and *C. lusitaniae* (lane 7). Lane M is a 100-bp DNA ladder and the positions of migration of 100- and 600-bp fragments are marked.

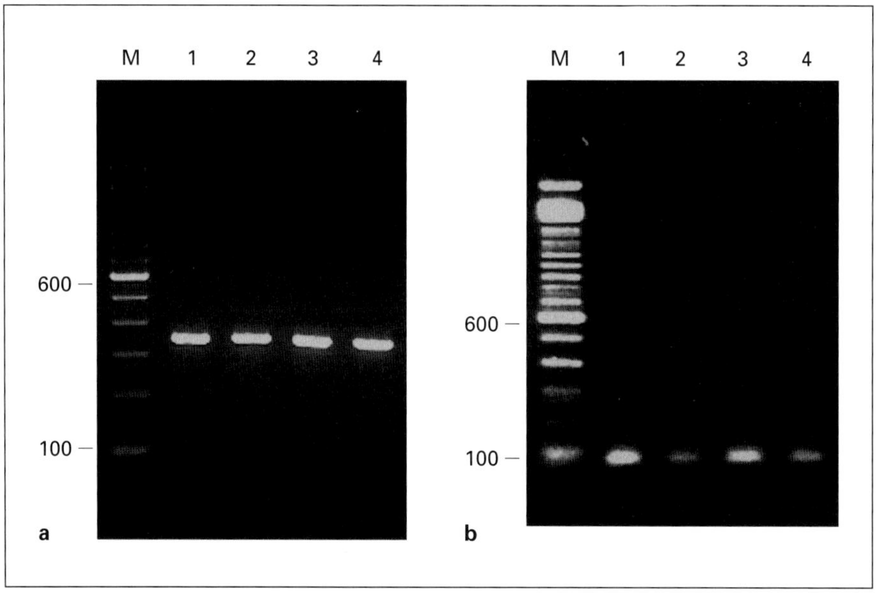

3

Fig. 3. Agarose gel electrophoresis of the product of the first round of PCR amplification using universal fungal primers (CTSF and CTSR) (**a**) and snPCR amplification using CTSR and CDDET primers (**b**) of genomic DNA from the 4 clinical *C. dubliniensis* isolates MF 1042, MF 41, MF 48 and MF 122 (lanes 1–4, respectively). Other details are the same as in the legend of figure 2.

other *Candida* species, i.e., *C. albicans, C. parapsilosis, C. tropicalis, C. glabrata, C. krusei*, and *C. lusitaniae* (fig. 2b), or other fungal pathogens (e.g. *A. fumigatus, F. solani, T. asahii* and *C. neoformans*; data not shown). No amplification was detected with genomic DNA from nonfungal organisms in the first as well as the second step of snPCR (data not shown).

The PCR amplification of rDNA from all the 12 clinical isolates of *C. dubliniensis* recovered from Kuwait, using CTSF and CTSR primers, resulted in specific amplification of a single DNA fragment of the expected size (the data from four isolates are shown in figure 2a, lanes 1–4) from each isolate. Further, reamplification of the products of the first PCR with CTSR and CDDET resulted again in specific amplification of a single DNA

Table 1. Phenotypic characteristics of *C. dubliniensis* isolates recovered from clinical specimens in Kuwait

Isolate No.	Source	Characteristics on SSA		Growth at or on				
		colony	chlamydo-spores	xylose	MDG	lac-tate	treha-lose	6.5% NaCl
MF 1042	urine	rough	+	–	–	–	–	–
MF 41	sputum	rough	+	–	–	–	–	–
MF 48	sputum	rough	+	–	–	–	–	–
MF 122	sputum	rough	+	–	–	–	–	–
S 4	sputum	rough	+	–	–	–	–	–
S 12	vaginal swab	rough	+	+	–	–	–	–
S 59	sputum	rough	+	+	–	–	–	–
MF 272	sputum	rough	+	–	–	–	–	–
MF 311	catheter tip	rough	+	–	–	–	–	–
MF 426	BAL	rough	+	+	–	–	–	–
MF 460	tracheal secretion	rough	+	+	–	–	–	–
MF 523	throat swab	rough	+	–	–	–	–	–

SSA = Sunflower seed agar; BAL = bronchoalveolar lavage; MDG = α-methyl-*D*-glucoside.

Table 2. Antifungal susceptibility data of 12 isolates of *C. dubliniensis* by E-test

Antifungal agents	Antifungal susceptibility, μg/ml			
	range	MIC_{50}	MIC_{90}	% resistance
Amphotericin B	0.002–0.38	0.023	0.047	0
Fluconazole	0.094–0.75	0.19	0.25	0
Voriconazole	0.003–0.012	0.006	0.012	0

fragment (the data from the same four isolates are shown in figure 2b, lanes 1–4) of the same size as that obtained from the reference strain of *C. dubliniensis.* Finally, the identity of two of the representative clinical isolates (MF 1042 and MF 122) as *C. dubliniensis* was confirmed by DNA sequencing of the ITS2 region of rDNA. The DNA sequence of the ITS2 region of these isolates completely matched only with the *C. dubliniensis* strain CD 36 but not with *C. albicans* or any other *Candida* species. These data confirmed the isolates recovered from HIV-negative patients in Kuwait as *C. dubliniensis.*

Discussion

This study reports recovery of 12 isolates of *C. dubliniensis* from a range of clinical samples and originating from patients with different underlying conditions within a span of 1 year. The identity of the isolates has been unequivocally established as *C. dubliniensis* by known phenotypic and molecular characteristics. In most of the published studies, differences in the assimilation of xylose, trehalose, and α-methyl-*D*-glucoside have been used to separate *C. dubliniensis* from *C. albicans* [1, 27, 28]. However, some recent reports suggest that assimilation characteristics of carbohydrates between the 2 species are not as discriminatory as initially reported and may vary depending upon the assimilation method employed [29, 30]. The ability of 4 of our *C. dubliniensis* isolates to assimilate *D*-xylose within 24 h with the Vitek 2 yeast identification system also suggests the inconsistency of the assimilation profiles in distinguishing these two closely related *Candida* species. Using a novel automated assimilation system, Kurzai et al. [30] reported that *C. dubliniensis* isolates were able to utilize *D*-xylose and trehalose. In another study [29], while the 24-hour profiles obtained with the Vitek 2 yeast identification system showed that α-methyl-*D*-glucoside, *D*-xylose and glycerol could be used to separate *C. dubliniensis* from *C. albicans,* at 48 h only *D*-xylose (6 vs. 95%) was useful for separating the two species. Following comparative evaluation of different commercial systems, these investigators [29] concluded that the assimilation of α-methyl-*D*-glucoside, *D*-xylose and trehalose was most useful for the differentiation of the two species by the majority of commercial systems. Moreover, it was the rapidity of the assimilation achieved with the commercial systems that allowed the differentiation of *C. dubliniensis* from *C. albicans.* It is apparent that while most of the *C. dubliniensis* isolates can be distinguished by carbohydrate assimilation results, some strains may give inconsistent results justifying application of PCR-based methods for unequivocal identification [29–32].

As our understanding on the occurrence and pathogenesis of this novel species is increasing, it is becoming apparent that this pathogen has the potential to cause a wide spectrum of clinical conditions both in HIV-infected and non-HIV-infected patients such as those receiving chemotherapy for cancer, organ transplant recipients or patients with other underlying conditions [2, 4, 6, 8, 33, 34]. A review of published reports suggests that *C. dubliniensis* is widely distributed in the population but forms only a minor constituent of the microbial flora in HIV-negative individuals [1, 2, 32]. Although the species may be found at any anatomic site, it probably exhibits greater propensity to colonize oral mucosa particularly of HIV-infected patients [2, 4, 5, 34, 35]. In our study, none of the *C. dubliniensis* isolates came from HIV-infected patients, but as many as 9 of the 12 (75%) were isolated from respiratory tract specimens (table 1). The other three isolates were recovered from catheter tip, urine and vaginal specimens. Other studies have also reported the isolation of *C. dubliniensis* from a wide range of clinical specimens including bronchoalveolar lavage, urine and catheter tip [37, 38]. In a study of the Irish population of normal healthy individuals, only 3.5% were found to carry *C. dubliniensis* in the oral cavity, while the prevalence of this species in the vagina was found to be even lower (2.7%) [34]. In contrast, 26% of HIV-infected and 32% of AIDS patients with symptoms of oropharyngeal candidiasis yielded *C. dubliniensis,* whereas in patients without symptoms of oropharyngeal candidiasis the colonization levels were 18 and 25%, respectively [35]. Like in other regions of the world [2, 32], occurrence of *C. dubliniensis* has also been recognized in the Middle East [20, 33, 34, 36–38]. Fotedar and Al-Hedaithy [38] studied 823 germ tube-positive and chlamydospore-positive *C. albicans* isolates at the University Hospital in Riyadh, Saudi Arabia. On reexamination, 27 (3.3%) of them were identified as *C. dubliniensis.* The prevalence of *C. dubliniensis* amongst germ tube-positive isolates was almost similar (3%) in the present study.

It has been suggested that the recent emergence of *C. dubliniensis* as a human pathogen may have resulted due to the widespread use of azole antifungal therapy in HIV-infected patients, in a manner similar to that suggested for *C. glabrata* [1, 2]. However, this assumption may not be correct since it has been found that the great majority of *C. dubliniensis* isolates are inherently susceptible to commonly used antifungal agents [5, 39]. Similarly, all the 12 *C. dubliniensis* isolates of this study were susceptible to fluconazole and voriconazole and also to amphotericin B.

Conclusions

We have described the isolation and identification of 12 isolates of *C. dubliniensis* recovered mostly from the respiratory tract of HIV-negative patients in Kuwait during a 1-year period. As our understanding of the epidemiology of *C. dubliniensis* is increasing, it is becoming apparent that this pathogen can colonize different anatomic sites with a potential to cause superficial or invasive disease.

Acknowledgements

We thank Rachel Chandy and Da'ad Farhat for excellent technical assistance. This work was supported by Kuwait University Research Administration grant No. MI 118.

References

1 Sullivan D, Coleman DC: *Candida dubliniensis:* Characteristics and identification. J Clin Microbiol 1998;36:329–334.
2 Sullivan DJ, Moran GP, Pinjon E, Al-Mosaid A, Stokes C, Vaughan C, Coleman DC: Comparison of the epidemiology, drug resistance mechanisms, and virulence of *Candida dubliniensis* and *Candida albicans.* FEMS Yeast Res 2004;4:369–376.
3 Odds FC, Van Nuffel L, Dams G: Prevalence of *Candida dubliniensis* isolates in a yeast stock collection. J Clin Microbiol 1998;36:2869–2873.
4 Jabra-Rizk MA, Falkler WA Jr, Merz WG, Baqui AA, Kelly JI, Meiller TF: Retrospective identification and characterization of *Candida dubliniensis* isolates among *Candida albicans* clinical laboratory isolates from human immunodeficiency virus (HIV)-infected and non-HIV-infected individuals. J Clin Microbiol 2000;38:2423–2426.
5 Moran GP, Sullivan DJ, Henman MC, McCreary CE, Harrington B, Shanley DB, Coleman DC: Antifungal drug susceptibilities of oral *Candida dubliniensis* isolates from human immunodeficiency virus (HIV)-infected and non-HIV-infected subjects and generation of stable fluconazole-resistant derivatives in vitro. Antimicrob Agents Chemother 1997;41:617–623.

6 Brandt ME, Harrison LH, Pass M, Sofair AN, Huie S, Li R-K, Morrison CJ, Warnock DW, Hajjeh RA: *Candida dubliniensis* fungemia: The first four cases in North America. Emerg Infect Dis 2000;6:46–49.
7 Meis JPGM, Ruhnke M, DePauw BE, Odds FC, Siegert W, Verweij PE: *Candida dubliniensis* candidemia in patients with chemotherapy-induced neutropenia and bone marrow transplantation. Emerg Infect Dis 1999;5:150–153.
8 Sebti A, Kiehn TE, Perlin D, Chaturvedi V, Wong M, Doney A, Park S, Sepkowitz KA: *Candida dubliniensis* at a cancer center. Clin Infect Dis 2001;32:1034–1038.
9 Coleman DC, Rinaldi MG, Haynes KA, Rex JH, Summerbell RC, Annaissie EA, Li A, Sullivan DJ: Importance of *Candida* species other than *Candida albicans* as opportunistic pathogen. Med Mycol 1998;36(suppl 1):156–165.
10 Khan ZU, Ahmad S, Mokaddas E, Chandy R: Simplified sunflower *(Halianthus annuus)* seed agar for differentiation of *Candida dubliniensis* from *Candida albicans.* Clin Microbiol Infect 2004;10:590–592.
11 Staib P, Morschhauser J: Chlamydospore formation on Staib agar as a species-specific characteristic of *Candida dubliniensis.* Mycoses 1999;42:521–524.
12 Al Mosaid A, Sullivan DJ, Coleman DC: Differentiation of *Candida dubliniensis* from *Candida albicans* on Pal's agar. J Clin Microbiol 2003;4:4787–4789.
13 Khan ZU, Ahmad S, Mokaddas E, Chandy R: Tobacco agar: A new medium for differentiation of *Candida dubliniensis* from *Candida albicans.* J Clin Microbiol 2004;42;4796–4798.
14 Mosca CO, Moragues MD, Llovo J, Al Mosaid A, Coleman DC, Ponton J: Casein agar: A useful medium for differentiating *Candida dubliniensis* from *Candida albicans.* J Clin Microbiol 2003;41:1259–1262.
15 Kirkpatrick WR, Revankar SG, Mcatee RK, Lopez-Ribot JL, Fothergill AW, McCarthy DI, Sanche SE, Cantu RA, Rinaldi MG, Patterson TF: Detection of *Candida dubliniensis* in oropharyngeal samples from human immunodeficiency virus-infected patients in North America by primary CHROMagar candida screening and susceptibility testing of isolates. J Clin Microbiol 1998;36:3007–3012.
16 Donnelly SM, Sullivan DJ, Shanley DB, Coleman DC: Phylogenetic analysis and rapid identification of *Candida dubliniensis* based on analysis of *ACT1* intron and exon sequences. Microbiology 1999;145:1871–1882.
17 Kurzai O, Heinz WJ, Sullivan DJ, Coleman DC, Frosch M, Muhlschlegel FA: Rapid PCR test for discriminating between *Candida albicans* and *Candida dubliniensis* isolates using primers derived from the pH regulated *PHR1* and *PHR2* genes of *C. albicans.* J Clin Microbiol 1999;37:1587–1590.
18 Ahmad S, Khan Z, Mustafa AS, Khan ZU: Seminested PCR for diagnosis of candidemia: Comparison with culture, antigen detection, and biochemical methods for species identification. J Clin Microbiol 2002;40:2483–2489.
19 Ahmad S, Mustafa AS, Khan Z, Al-Rifaiy AI, Khan ZU: PCR-enzyme immunoassay of rDNA in the diagnosis of candidemia and comparison with amplicon detection by agarose gel electrophoresis. Int J Med Microbiol 2004; 294:45–51.
20 Ahmad S, Khan Z, Mokaddas E, Khan ZU: Isolation and molecular identification of *Candida dubliniensis* from non-human immunodeficiency virus-infected patients in Kuwait. J Med Microbiol 2004;53:633–637.
21 McGinnis MR: Mycology; in Isenberg HD (ed): Clinical Microbiology Procedures Handbook. Washington, American Society for Microbiology Press, 1994, section 6, pp 6.1.1.–6.1.12.
22 Alves SH, Milan EP, de Laet Sant'Ana P, Oliveira LO, Santurio JM, Colombo AL: Hypertonic sabouraud broth as a simple and powerful test for *Candida dubliniensis* screening. Diagn Mirobiol Infect Dis 2002;43:85–86.
23 Ahmad S, Mokaddas E, Al-Anoud J: Rapid detection of ethambutol-resistant *Mycobacterium tuberculosis* strains by PCR-RFLP targeting *embB* codons 306 and 497 and *iniA* 501 mutations. Mol Cell Probes 2004;18:299–306.
24 National Committee for Clinical Laboratory Standards: Reference Method for Broth Dilution Antifungal Susceptibility Testing of Yeast, Approved Standard M27-A. Wayne, National Committee for Clinical Laboratory Standards, 1997.
25 Pfaller MA, Messer SA, Karlsson A, Bolmstrom A: Evaluation of the Etest method for determining fluconazole susceptibilities of 402 clinical yeast isolates by using three different agar media. J Clin Microbiol 1998;36:2586–2589.
26 Pfaller MA, Messer SA, Houston A, Mills K, Bolmstrom A, Jones RN: Evaluation of the Etest method for determining voriconazole susceptibilities of 312 clinical isolates of *Candida* species by using three different agar media. J Clin Microbiol 2000;38:3715–3717.
27 Gales AC, Pfaller MA, Houston AK, Jolly SDJ, Sullivan DJ, Coleman DC, Soll DR: Identification of *Candida dubliniensis* based on temperature and utilization of xylose and alpha-methyl-*D*-glucoside as determined with the API 20C AUX and Vitek YBC Systems. J Clin Microbiol 1999;37:3804–3808.
28 Sancak B, Rex JH, Paetznick V, Chen E, Rodriguez J: Evaluation of a method for identification of *Candida dubliniensis* bloodstream isolates. J Clin Microbiol 2003;41:489–491.
29 Pincus DH, Coleman DC, Pruitt WR, Padhye AA, Salkin IF, Geimer M, Bassel A, Sullivan DJ, Clarke M, Hearn V: Rapid identification of *Candida dubliniensis* with commercial yeast identification systems. J Clin Microbiol 1999; 37:3533–3539.
30 Kurzai O, Korting HC, Harmsen D, Bautsch W, Molitor M, Frosch M, Muhlschlegel FA: Molecular and phenotypic identification of the yeast pathogen *Candida dubliniensis.* J Mol Med 2000;78:521–529.
31 Tintelnot K, Haase G, Seibold M, Bergmann F, Staemmler M, Franz T, Naumann D: Evaluation of phenotypic markers for selection and identification of *Candida dubliniensis.* J Clin Microbiol 2000;38:1599–1608.
32 Gutierrez J, Morales P, Gonzalez MA, Quindos G: *Candida dubliniensis,* a new fungal pathogen. J Basic Microbiol 2002;42:207–227.
33 Al-Hedaithy SS: The yeast species causing fungemia at a university hospital in Riyadh, Saudi Arabia, during a 10-year period. Mycoses 2003;46:293–298.
34 Polacheck I, Strahilevitz J, Sullivan D, Donnelly S, Salkin IF, Coleman DC: Recovery of *Candida dubliniensis* from non-human immunodeficiency virus-infected patients in Israel. J Clin Microbiol 2000;38:170–174.
35 Ponton J, Ruchel R, Clemons KV, Coleman DC, Grillot R, Guarro J, Aldebert D, Ambroise-Thomas P, Cano J, Carrillo-Munoz AJ, Gene J, Pinel C, Stevens DA, Sullivan DJ: Emerging pathogens. Med Mycol 2000;38 (suppl 1):225–236.
36 Lefler E, McCullough MJ, Clemons KV, Stevens DA: Initial isolation of *Candida dubliniensis* from the Middle East. Int J Infect Dis 2001;5:40–42.
37 Fotedar R, Al Hedaithy SS: Prevalence of *Candida dubliniensis* among germ tube-positive yeasts recovered from the respiratory specimens in HIV-negative patients. Mycoses 2004; 47:50–155.
38 Fotedar R, Al-Hedaithy SS: *Candida dubliniensis* at a university hospital in Saudi Arabia. J Clin Microbiol 2003;41:1907–1911.
39 Quindos G, Carrillo-Munoz AJ, Arevalo MP, Salgado J, Alonso-Vargas R, Rodrigo JM, Ruesga MT, Valverde A, Peman J, Canton E, Martin-Mazuelos E, Ponton J: In vitro susceptibility of *Candida dubliniensis* to current and new antifungal agents. Chemotherapy 2000; 46:395–401.

Original Paper

Medical Principles and Practice

Med Princ Pract 2005;14(suppl 1):84–91
DOI: 10.1159/000086189

Received: October 27, 2004
Revised: February 16, 2005

Instability of Chimaeric Antibody Secretion by Anti-Carcinoembryonic Antigen Producing Hybridoma Cells after Gene Targeting

Abir A. El-Abdallah[a] Christopher H.J. Ford[b]

Departments of [a]Pathology and [b]Surgery, Faculty of Medicine, Kuwait University, Kuwait

Key Words

Chimaeric antibody · Gene targeting · Homologous recombination

Abstract

Objective: To produce a chimaeric version of the 11-285-14 anti-carcinoembryonic antigen (CEA) monoclonal antibody using a gene targeting approach. ***Materials and Methods:*** A replacement vector was constructed to insert the human constant $\gamma 1$ gene within the mouse heavy chain locus of 11-285-14 hybridoma cells. The mouse constant $\gamma 1$ gene (1.5 kb) and the mouse μ intron fragment (2.2 kb) were amplified by PCR and cloned into a pKO Scrambler vector. The human constant $\gamma 1$ gene fragment (2.2 kb) was cloned next to the intron fragment. Resistant colonies were screened by ELISA for the presence of the human isotype in their supernatants. ***Results:*** Of the 4,370 resistant colonies obtained, 87 colonies showed secretion of the human isotype at levels between 4 and 32 ng/ml. PCR and Southern blot results confirmed the correct integration of the human gene by homologous recombination within the heavy chain locus. Most of the producers ceased to express the human isotype within a few weeks after the initial positive ELISA results. Instability of secretion could not be explained by genetic instability in all the clones, which suggests the presence of other undefined epigenetic or physiologic mechanisms. ***Conclusion:*** Gene targeting resulted in transformants with unstable and low production rates of chimaeric anti-CEA antibody.

Introduction

Therapeutic monoclonal antibodies (mAbs) have established themselves as important additional tools in the management of many diseases including cancer. Many mAbs are FDA approved for human use [1]; half of them are humanized by chimaerization (fusion of the variable region of the mouse mAb and the constant region of a human antibody). Clinical trials with chimaeric antibodies have shown that they are successful therapeutic agents because the insertion of human constant regions improves interaction with human Fc receptors, reduces immunogenicity and extends the metabolic survival of the therapeutic antibody [2].

11-285-14 is a mouse anti-human carcinoembryonic antigen (CEA) antibody of the immunoglobulin G1 (IgG1) isotype. This antibody has high specificity to CEA and does not cross-react with non-specific cross-reacting antigen. 11-285-24 and its derivatives have shown promising therapeutic effects with CEA expressing tumours in vitro and in preclinical models in vivo [3–5]. Significant differences exist in the ability of therapeutic antibodies

KARGER

Fax +41 61 306 12 34
E-Mail karger@karger.ch
www.karger.com

1011–7571/05/0147–0084$22.00/0

Accessible online at:
www.karger.com/mpp

Dr. Abir A. El-Abdallah
Department of Pathology, Faculty of Medicine, Kuwait University
PO Box 24923
13110 Safat (Kuwait)
Tel. +965 5312300/6250, Fax +965 5338905, E-Mail abeer.a@hsc.edu.kw

to kill cancer cells even when these antibodies recognize the same antigen. These differences cannot be identified without extensive in vivo evaluation [1]. This is hampered mainly by the development of human anti-mouse antibodies (HAMAs) following repeated administration of antibody, which can cause deleterious effects to the patient and interfere with the therapeutic activity of the antibody. HAMAs are directed mainly against the Fc portion of the therapeutic antibody. Studies have shown that HAMA responses are greater and faster to whole Ig than to the divalent or monovalent fragments, consistent with the concept that the presence of the Fc portion increases the immunogenicity of Ig [2]. For 11-285-14 to be preclinically evaluated, a less immunogenic version is needed. Hence, the objective of this study was to produce a chimaeric version of 11-285-14 antibody by gene targeting the Ig heavy chain locus of the 11-285-14 hybridoma cells and inserting the human constant gene therein.

The choice of a gene targeting approach was based on the concept that homologous recombination allows the introduced human γ1 gene to be expressed in a normal chromosomal context within the mouse heavy chain locus. The anticipated advantage of this approach was the stable and the high Ig expression reported to be obtained from targeted hybridomas [6]. If this could be accomplished, the problem of insufficient chimaeric antibody production rates and instability often seen in conventional transfectomas would be circumvented [7].

Materials and Methods

Construction of Targeting Vector

RNA and DNA were extracted from 11-285-14 hybridoma cells by TRIzol (Gibco-BRL). RNA was reverse transcribed to cDNA using the first-strand cDNA synthesis kit (Amersham Pharmacia Biotech). The homology regions were amplified by PCR using primers (Oswel) designed based on sequences available in GenBank and the MasterAmp High Fidelity Long PCR Kit (Epicentre Technologies). PCR products were prepared for cloning into the pKO-NTK vector (Stratagene) by digestion with compatible restriction enzymes (Pharmacia).

The 1.5-kb mouse constant gene was digested with *Bam*HI and *Eco*RI and the 2.2-kb intron fragment was digested with *Bgl*II and *Hin*dIII. The 2.2-kb human constant gene was prepared for cloning next to the intron fragment by digestion with *Hin*dIII at the 5′ end and creating a blunt 3′ end by cutting with *Sp*HI and treating with Klenow enzyme. The second targeting vector, having the functionally rearranged VDJ region as the 5′ homology region, was prepared in similar a way. The VDJ region was amplified by RT-PCR, sequenced and then digested with *Eco*RV and *Hin*dIII and cloned in frame to the human constant gene. Preparation and cloning were done according to standard procedures. Proper construction of the two vectors was confirmed by sequencing.

Transfection of Hybridoma Cells and Selection of Homologous Recombinants

The targeting vector was linearized and transfected into 11-285-24 hybridoma cells by electroporation at 0.23 kV, 500 μF at room temperature. The quantity of DNA used for transfection varied from 1 to 12 μg. Selection of recombinant cells was based on the activity of the positive selection marker (NEO) that confers resistance to G418, and the negative selection marker (TK) that confers resistance to ganciclovir (GANC). Optimal concentration of G418 was determined by titration (0.5 mg/ml). GANC was used at a concentration of 2 *μM*. After electroporation, cells were cultured in bulk for 2 days then plated at 10^5 cells/well in 24-well plates. Homologous recombinant cells were selected by 1 week of culture in G418 and 1 week in GANC. Supernatants were screened by enzyme-linked immunosorbent assay (ELISA) for secretion of the human isotype.

Enzyme-Linked Immunosorbent Assay

Plates were coated with 10 μg/ml of rabbit anti-human IgG (Dako), blocked with normal rabbit serum and 10% fetal calf serum in carbonate, and the supernatants were added. Rabbit anti-human IgG conjugated to horseradish peroxidase (HRP; Dako), diluted 1/1,000 in 10% fetal calf serum in PBS, was used as detection reagent. The signal was developed by addition of the substrate 2,2-azino-di-3-ethyl benzothiazoline sulphonic acid that generates a coloured reaction product measured spectrophotometrically at 405 nm. ELISA readings were compared to those obtained from 11-285-14 parent cell supernatants and human IgG1 standards. A positive result was operationally defined as a reading greater than the mean +2 standard deviations of the value obtained for parent cell supernatant tested in triplicate.

Southern and Western Blots

Southern blot experiments were performed by the alkali transfer method according to standard procedures. The human probe was labelled with alkaline phosphatase direct DNA probe labelling kit (Amersham Pharmacia Biotech) and used in conjunction with a chemiluminescent detection system. Western blot experiments were done according to standard procedures. Immunodetection was done using an anti-human Ig antibody conjugated to HRP (Dako) and ECL Plus Western blotting detection reagents (Amersham Biosciences). The molecular weight of the bands was determined according to a molecular weight marker curve, drawn using the Excel program.

Results

Functionally Rearranged Allele

To draw a complete picture of the 11-285-14 functionally rearranged heavy chain locus, the VDJ fragment was amplified by RT-PCR from 11-285-14 RNA, cloned into pGEMT vector and sequenced. Two different VDJ regions could be amplified using two degenerate primers that anneal to two different families of variable genes. The reverse primer was a degenerate primer that anneals to the beginning of the mouse constant γ genes. This

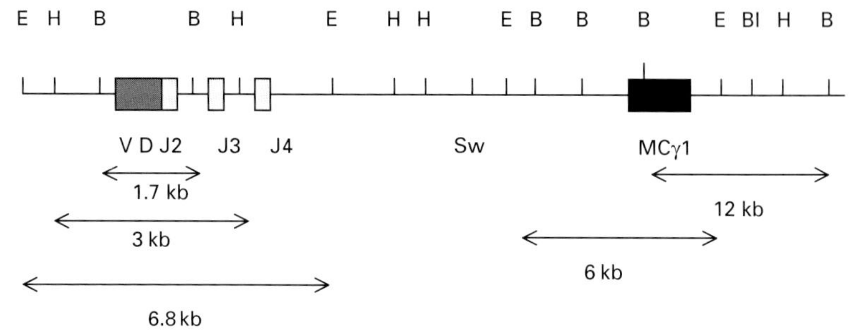

Fig. 1. Genomic situation of the functionally rearranged heavy chain locus in the hybridoma 11-285-14. Sw is the start of the switch region. Restriction sites: B = *Bam*HI; E = *Eco*RI; H = *Hin*dIII; Bl = *Bgl*II. Diagram is not drawn to scale.

showed that 11-285-14 contained two variable genes that are rearranged, class switched to a constant γ gene, transcribed and expressed at the mRNA level. The sequences of the two fragments were analysed using the International Immunogenetics Information System (http://imgt.cines.fr). One of the VDJ regions (V5) matched the IgHV5S12 of the Vh 7183 family of germ line genes in record U04228. Its diversity segment was identified as D-FL16.1, and its joining segment was identified as J2. V5 variable region was also found to be homologous to the mouse anti-CEA heavy chain variable gene published in GenBank (GenBank: gi 5708432). The other VDJ region matched the IgHV14S1 germ line variable gene. Its diversity region could not be identified and its joining segment matched in part J2. It partly matched J4 as well. A stop codon (TAG) was detected within its diversity region at amino acid number 105. The presence of the stop codon along with the unclear picture of the diversity and J regions led to the consideration that V4 is the aberrant allele of 11-285-14 and V5 is the functionally rearranged allele. Based on these facts, the target Ig heavy chain locus was predicted to be as shown in figure 1.

Construction of Targeting Vectors

A replacement vector was constructed to target the Ig heavy chain locus of the 11-285-14 hybridoma cells and to insert therein a human constant gene (fig. 2a). The 5′ and 3′ homology regions were isogenic DNA fragments amplified by PCR and cloned into the pKO vector that had a positive-negative selection system built in. A positive selection marker (NEO) was placed between the two regions of homology; this confers resistance to G418 drug. A negative selection marker [herpes simplex virus thymidine kinase (HSV-TK)] was placed outside the region of homology; this gives the cells the ability to metabolize GANC, leading to the formation of a toxic metabolite. Homologous recombination between the targeting vector and genomic DNA results in the loss of the HSV-TK sequences; such cells are selected by being resistant to G418 and GANC. On the other hand, non-homologous insertion of exogenous DNA into the genome results in the persistence of the HSV-TK sequence; such cells are resistant to G418 but sensitive to GANC. The targeting strategy is shown in figure 2b.

Secretion of Chimaeric Antibody

Forty-eight electroporation experiments were performed, which resulted in 4,370 resistant colonies. Eighty-seven of these colonies showed secretion of the human isotype as determined by positive readings in ELISA (table 1). The level of the human isotype detected was in the range of 4–32 ng/ml of supernatant. Colonies in wells with positive readings were selected and cultured individually. Cells were left to grow to confluency between 3 and 7 days and a second ELISA test was performed. Some of these colonies continued to show positive ELISA readings, while others became negative. With continuous culturing, all colonies became negative in ELISA. Also, some of the resistant colonies died in long-term culture. All the clones initially identified as human isotype secretors continued to secrete the murine anti-CEA antibody.

Integration of the Human γ1 Gene by Homologous Recombination

Representative clones were selected from transformants that had been identified as chimaeric antibody secretors by ELISA and lost secretion in continuous culture. Parent hybridoma cells were used as a negative control. A 500-bp fragment of the human constant gene could be amplified by PCR performed using DNA extracted from these clones (fig. 3a). To confirm the identity of the am-

Fig. 2. Schematic diagrams of the targeting vector and the targeting strategy. **a** The targeting construct is a replacement vector that contains the human IgG1 constant gene (HuCγ1) flanked by sequences homologous to the mouse μ intron at the left-hand side and sequences homologous to the mouse IgG1 constant gene at the right-hand side. Two selection markers, neomycin (NEO) and thymidine kinase (TK), are incorporated within the constructs to allow enrichment for the homologous recombinant cells. Arrows in the vector diagram indicate the primers used for sequencing. Restriction enzyme sites show the sites of linearization of the vector. **b** Targeting strategy for the development of a chimaeric antibody by targeting the mouse Ig heavy chain locus with the replacement vector. The predicted chimaeric situation at the 11-285-14 heavy chain locus induced by double cross-over at the homologous sequences is shown. The 6.4-kb *Bam*HI fragment hybridizing to the human gene probe is diagnostic of integration at the correct place. The 11-kb *Bgl*II fragment is another diagnostic fragment. This picture applies for replacement events within the functional and the non-functional heavy chain alleles. The only difference is the distribution of the restriction sites before the VDJ regions. The non-functional VDJ lies within a 2.7-kb *Bam*HI fragment instead of a 1.7-kb *Bam*HI fragment for the functional allele.

Fig. 3. a Amplification of a fragment of the human γ1 gene from DNA extracted from resistant transfectomas (lanes 2–4). The amplified product is a 500-bp fragment. Lane 1 is parent cell DNA used as a negative control. Lane 5 is the 100-bp ladder. **b** Southern blotting experiment using the human constant γ1 gene as a probe. DNA was extracted from colonies that were initially positive in ELISA and had lost secretion after culture. DNA was digested with *Bam*HI (gel 1). Lane 1 is parent cell DNA. Lane 2 is human white blood cell DNA showing the expected bands of 7.5 and 11 kb in size as calculated by analysing the sequence of the human Ig heavy chain locus available in GenBank. The expected 6.4-kb band indicating integration by homologous recombination was detected in sample No. 3 (lane 5). Additional bands of 9 and 14 kb were detected in the other samples possibly due to partial digestion or integration in other alleles. Gel 2 shows results of the same samples after cutting the DNA with *Bgl*II. The expected 11-kb band is seen in sample No. 3 (lane 3).

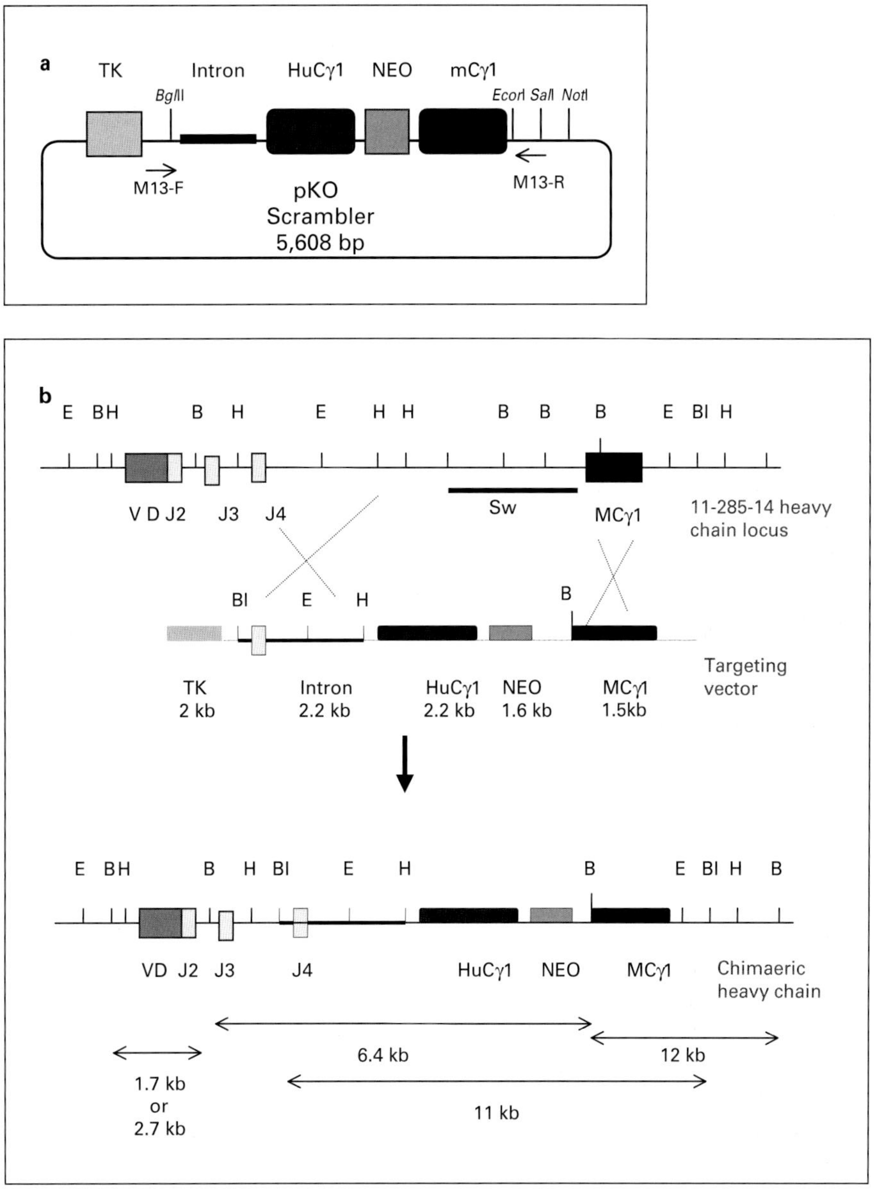

2

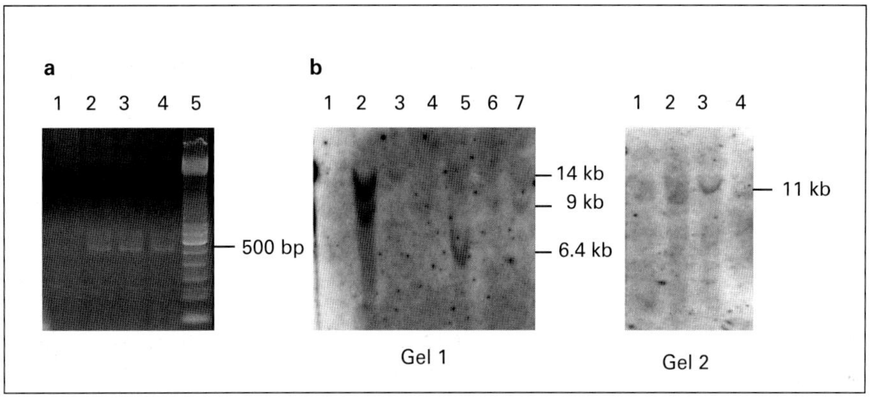

3

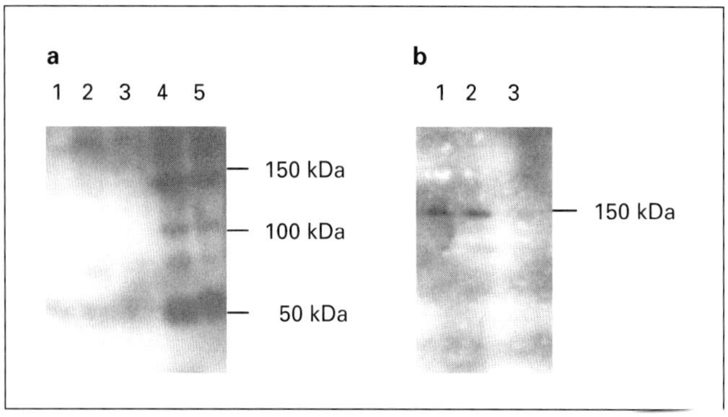

Fig. 4. Western blotting and immunodetection performed on protein extracted from resistant clones identified as chimaeric antibody secretors by ELISA. All samples were prepared under non-reducing conditions and run on non-reducing 0.5% SDS acrylamide gels. Immunodetection was done with HRP anti-human Ig (Dako). **a** Lane 1 is protein extracted from parent cells. Lanes 2–5 are protein extracted from samples described above. 150 kDa is the size of a completely assembled antibody, while 50 kDa is the size of the human heavy chain; 100 kDa is a heavy chain dimer. **b** The samples in lanes 1 and 2 were protein extracted from 12D1G7 colony that showed positive results in Southern blot experiment (fig. 3). This colony was plated by semi-limiting dilution and the subclones gave positive readings in second and third ELISAs. The sample in lane 3 in the same figure is protein extracted from a colony (12A1) that was positive in the first ELISA screening and lost the ability to grow in selection medium with continuous culture (cells have lost the NEO gene and most probably the human gene also); these cells were subsequently grown in normal medium in order to get enough cells for protein extraction.

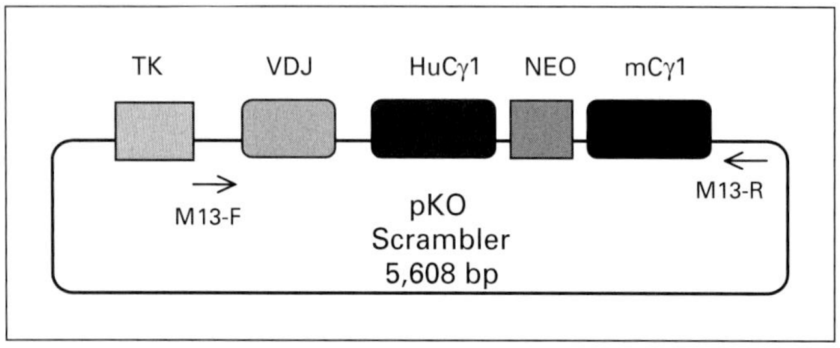

Fig. 5. Schematic diagram of the second targeting vector having the functionally rearranged VDJ region as the 5′ homology region cloned in frame to the human constant gene.

plified DNA, the PCR product was sequenced (sequences not shown). The sequences obtained were compared to the human and mouse genomes using the GenBank Blast program. Comparison showed 100% homology with sequences on human chromosome 14 (human IgG) and no similarity with the mouse genome. According to the prediction, Southern blotting of DNA extracted from clones that had stopped secretion of the human isotype, after digestion with *Bam*HI or *Bgl*II, resulted in the detection of the expected 6.4-kb-sized and the 11-kb-sized fragments (fig. 2b) in one of the four representative transformants (fig. 3b). These results provide evidence that the human gene is still integrated within the heavy chain locus of this clone, even after losing secretion of the human isotype.

Screening for the Presence of the Human Isotype Intracellularly

The intracellular pool of the synthesized chimaeric heavy chains was investigated by Western blotting, performed on proteins extracted from the resistant colonies which were initially positive in ELISA. Most of these experiments gave negative results because by the time (almost 2 weeks) these colonies reached a sufficient number of cells to enable extraction of enough protein for the assay, they were losing the ability to secrete the human isotype. In two of the Western blot trials, the presence of the anticipated chimaeric antibody was shown (fig. 4a, b).

Trials to Stabilize and Enhance Secretion of the Chimaeric Antibody

To increase the chance of targeting to the functional gene locus, a second targeting vector was constructed, which had the functionally rearranged VDJ region (V5) as the 5′ homology arm. This was cloned in frame with the human constant gene within the pKO vector that already had the mouse constant gene as the 3′ homology arm (fig. 5). The results of electroporations with this vector were similar to those with the intron vector (table 1). All the positive clones lost secretion of the human isotype within few weeks of the initial positive ELISA. Different linearization sites for the two targeting vectors were also tried but this did not affect or enhance the stability of secretion. Since methylation of the human transgene can be one of the reasons why colonies initially identified as secretors stop secreting the chimaeric antibody, the anti-methylating agent – azacytidine – was used at a sublethal dose (4 μ*M*) to inhibit methylation. No effect on expression of the human isotype was detected. In another attempt to enhance the secretion of chimaeric antibody, cells were cultured in medium containing IL-6 (1 ng/ml) since it is one of the cytokines known to increase the secretion of antibodies in hybridomas. After 1 week of culture, the supernatants were tested by ELISA. No human isotype was detected.

Table 1. Summary of electroporation experiments that resulted in secreting colonies

Vector	Quantity of DNA μg	Linearization site	Transfection efficiency[a]	Resistant colonies[b]	Positive wells[c]	Homologous recombination frequency[d]	Absolute frequency[e]
Intron	1	*Not*I	45	45	3	6.6	1.5
Intron	1	*Not*I	106	548	15	2.7	7.5
Intron	2	*Not*I	150	580	10	1.7	5
Intron	5	*Bgl*II/*Sal*I	19	187[f]	5	2.7	0.5
Intron	5	*Not*I/*Sal*I	10	115	2	1.7	0.2
Intron/VDJ	2/2	*Not*I/*Sal*I	73	146	8	5.5	1.6
Intron	2	*Not*I	100	209	11	5.3	2.2
Intron	7	*Bgl*II/*Sal*I	21	211[f]	6	2.8	0.6
Intron	5	*Not*I	12	116[f]	14	12.0	1.4
VDJ	6	*Eco*RI	39	390[f]	13	3.3	1.3

[a] The number of G418-resistant colonies per 10^6 plated cells.

[b] Total number of colonies resistant to G418 and GANC.

[c] Number of wells containing resistant colonies with supernatants that were positive in ELISA.

[d] Homologous recombination frequency (percentage) was calculated by dividing the total number of positive colonies by the total number of $G418^r$ $GANC^r$ colonies × 100. Considering one well contains one secreting colony.

[e] Absolute frequency is an estimate of homologous recombination events per 10^6 transfected cells. It is calculated by dividing the total number of positive colonies by the number of transfected cells × 10^6.

[f] Selection was with G418 only.

Discussion

The targeting vectors constructed to introduce the human constant γ1 gene into the 11-285-14 heavy chain locus were replacement-type vectors (fig. 2, 5). Studies have revealed that chromososmal Ig μ genes modified by gene replacement (double reciprocal recombination) are expressed at wild-type levels, whereas those modified by vector integration (single reciprocal recombination) have lower levels of Ig μ gene expression [8]. Replacement vectors also resulted in a consistently higher frequency of homologous recombination than integration vectors with less variation between cell lines [9]. It is known that gene targeting with an integration vector yields a recombinant structure in which the introduced DNA is flanked by duplicate DNA segments. Secondary recombination between these duplicated homologous regions may occur and thus reverses the targeting. Recombinants with integration events in the Ig heavy chain locus were found to be especially unstable [10]. It has been shown that homologous recombination in hybridoma cells is similar to that observed in other mammalian cell types (10^{-6} recombinants/cell), while random recombination is greater than 1,000-fold more frequent, and most transfected cells are not even targeted [11]. The necessity for a selection system that enriches homologous recombinant cells and the possibility of application of the positive-negative selection system only to replacement vectors rendered this type of vector more acceptable than integration vectors. All of this suggested that gene replacement may be a preferred method for the construction of hybridomas producing chimaeric antibodies.

The design of the targeting vector was based on information available in the literature [9–12]. The 5′ homology was chosen to be an isogenic DNA fragment of the mouse VDJ-Cμ intron that contains the enhancer and the matrix attachment regions necessary to maintain stable expression of the constant genes [13]. The 3′ homology region was 1.5 kb of isogenic DNA from the mouse constant γ1 gene. Since 11-285-14 hybridoma secretes anti-CEA antibody of IgG1 isotype, it was expected that the functional heavy chain gene locus (the target) is class switched to the constant γ1 gene (fig. 1). The transgene was a 2.2-kb fragment of the human constant gene that starts with the 3′ end of the intron before the CH1 exon and ends after the poly A termination signal. Therefore, production of the chimaeric antibody can only occur if the human constant gene is integrated downstream to an appropriate transcriptional promoter, translation initiation site, and splice donor sequence provided by the endogenous functionally rearranged variable gene segment. Positive reading in ELISA screening for secretion of the human isotype is indicative of integration of the human gene by homologous recombination within the Ig heavy chain locus.

The homologous recombination frequencies obtained in this study (table 1) were higher than those reported for other comparable integration vectors (0.46% in Kardinal et al. [10] and 0.5% in Mocikat et al. [6]) or replacement vectors (0.35–1% in Yarnold and Fell [9]; 0.99–5.69% in Kardinal et al. [12]) targeted to an isogenic Ig heavy chain locus. The reason may be the different cell line used in this study since it was shown that the efficiency of mediating homologous recombination process could vary significantly among cell lines [9]. Although the homologous recombination events were obtained at a high enough efficiency, the human isotype was expressed at low levels in the positive clones. This was indicated by the ELISA results showing a level of secretion in the range of 4–32 ng/ml. The results in all the electroporations that gave positive clones followed the same pattern of initial positivity, followed by negative readings after subcloning. These results suggested that some changes were occurring in the cells during selection and cultivation.

The chromosomal IgH locus is a hot spot for intrachromosomal homologous recombination, promoted by the enhanced transcriptional activity and the recombinationally active chromatin environment within this locus [14]. The ability of most of the clones to grow normally in selection medium during long-term culture indicated that they still had the NEO gene. The stable integration of the human γ1 gene was proved by PCR amplification of a 500-bp fragment of the human gene from DNA from positive clones after long-term culture (2–3 months). Southern blot results provided evidence that the human gene remained within the heavy chain locus in some of the clones and was not excised out by secondary recombination, even after losing secretion. We cannot exclude the possibility that the human gene could be excised from some of these clones but not all of them during culture. The small proportion of cells still containing the human gene secrete the human isotype at a low level undetected by the ELISA used for screening. Also, it could be that DNA from these cells showed the diagnostic fragments with the human probe on Southern blotting. This low level of target gene (low number of cells still having the human gene) could explain the low intensity of bands obtained in Southern blot although a lot of DNA and a highly sensitive probe were used.

Fundamental understanding of changes during the processes of selection of recombinant clones and their adaptation to the subcloning growth environment is lacking. Some cells may undergo irreversible alteration at the genome level; some may exhibit changes in their gene expression pattern, while others may acquire modulation of various fluxes by changing nutrient/metabolite concentrations and enzyme activities [15]. Gene integration, transcription, translation, protein processing and the secretion apparatus all affect the expression level of a gene product. In antibody-producing cells in particular, stable productivity is a complex issue, since separate genes, which are regulated independently, encode heavy and light chains. Secretion of the mature antibody into the supernatant depends on accurate assembly of the chains in the endoplasmic reticulum [16].

Stability of gene expression in recombinant antibody-producing cells is described by several groups with differing results; some reported stable expression while others reported loss of gene expression during long-term culture (summarized in Borth et al. [17]). Producers do convert to non-producers by genetic rearrangement, regulation of expression or improper posttranslational processing. Even at low conversion rates of producer to non-producer, the non-producer would have an inherent growth advantage, which would eventually cause them to outgrow the producers [18]. One group reported loss of heavy chain as the main cause of loss of antibody production and non-producing subclones appeared after only 2 weeks of culturing [19]. Others have reported a continuous decrease in light and heavy chain production during long-term culture without the appearance of a non-producing subpopulation. This indicated that the assembly in the endoplasmic reticulum is one of the major rate-limiting factors in antibody production [17]. It is worth noting that 11-285-14 hybridoma cells have shown high stability in long-term culture with loss of secretion occurring very infrequently. In addition, targeted clones that had lost secretion of the human isotype continued to secrete the mouse isotype after long-term culture.

Although there have been many reports on the production kinetics of mouse hybridoma and transfectoma cell lines [20–23], none, to our knowledge, has analysed the changes that may happen during the selection and adaptation process of gene-targeted hybridoma cells. The few published papers on production of chimaeric antibody by gene targeting showed this approach to be highly efficient [8–10]. The only problem reported in those studies, which dealt with interrupted secretion, was due to secondary recombination caused by duplication of sequences, since the targeting vectors used there were insertion vectors [24, 25].

We report here our experience of using gene targeting for the development of a chimaeric version of the 11-285-14 anti-CEA antibody. Although the homologous recombination efficiency obtained in this study was comparable and even better than in other similar studies, no clones producing chimaeric antibody at the wild-type level were obtained. This questions the efficiency of application of this method to all types of cell lines. We have used here replacement vectors and have shown that loss of the human transgene is not the cause of losing the human isotype secretion in all cases. Instability of secretion was not due to secondary recombination but to other yet undefined epigenetic or physiologic mechanisms that will form the basis of interesting areas of research for the future.

Conclusion

Gene targeting resulted in transformants with unstable and low production rates of chimaeric anti-CEA antibody.

Acknowledgment

This work was supported by the College of Graduate Studies and the Research Administration (grant No. MS 047), Kuwait University.

References

1 Glennie MJ, van de Winkel JGJ: Renaissance of cancer therapeutic antibodies. Drug Discov Today 2003;8:503–510.
2 Clark M: Antibody humanization: A case of the emperor's new clothes? Immunol Today 2000;21:397–402.
3 Ford CHJ, Bartlett SE, Casson AG, Marsden CH, Gallant ME: Efficacy and specificity of vindesine monoclonal anti-carcinoembryonic antigen conjugates with nine human cancer cell lines. NCI Monogr 1987;3:107–116.
4 Casson AG, Ford CHJ, Marsden CH, Gallant ME, Bartlett SE: Efficacy and selectivity of vindesine monoclonal anti-CEA antibody conjugates on human tumour cell lines grown as xenografts in nude mice. NCI Monogr 1987;3: 117–124.
5 Ford CHJ, Osborne PA, Rego BG, Mathew A: Bispecific antibody targeting of doxorubicin to CEA-expressing colon cancer cell lines in vitro and in vivo. Int J Cancer 2001;92:851–855.
6 Mocikat R, Kardinal C, Lang P, Zeidler R, Thierfelder S: Unaltered immunoglobulin expression in hybridoma cells modified by targeting of the heavy chain locus with an integration vector. Immunology 1995;84:159–163.
7 Kim JH, Bae SW, Hong HJ, Lee GM: Decreased chimaeric antibody productivity of KR12H-1 transfectoma during long-term culture results from decreased antibody gene copy number. Biotechnol Bioeng 1996;51:479–487.
8 Baker MD, Karn HA, Read LR: Restoration of a normal level of immunoglobulin production in a hybridoma cell line following modification of the chromosomal immunoglobulin μ gene by gene replacement. J Immunol Methods 1994;168:25–32.
9 Yarnold S, Fell HP: Chimerization of antitumour antibodies via homologous recombination conversion vectors. Cancer Res 1994;54: 506–512.
10 Kardinal C, Hooijberg E, Lang P, Zeidler R, Mocikat R: Integration vectors for antibody chimerization by homologous recombination in hybridoma cells. Eur J Immunol 1995;25: 792–797.
11 Vasquez KM, Marburger K, Intody Z, Wilson JH: Manipulating the mammalian genome by homologous recombination. Proc Natl Acad Sci USA 2001;98:8403–8410.
12 Kardinal C, Selmayr M, Mocikat R: Genetic stability of gene targeted immunoglobulin loci. 1. Heavy chain isotype exchange induced by a universal gene replacement vector. Immunology 1996;89:309–315.
13 Ronai D, Berru M, Shulman MJ: Variegated expression of the endogenous immunoglobulin heavy-chain gene in the absence of the intronic locus control region. Mol Cell Biol 1999;19: 7031–7040.
14 Raynard SJ, Read LR, Baker MD: Evidence for the murine IgH locus acting as a hot spot for intrachromosomal homologous recombination. J Immunol 2002;168:2332–2339.
15 Korke R, Rink A, Seow TK, Chung MCM, Beattie CW, Hu WS: Genomic and proteomic perspectives in cell culture engineering. J Biotechnol 2002;94:73–92.
16 Strutzenberger K, Borth N, Kunert R, Steinfellner W, Katinger H: Changes during subclone development and ageing of human antibody-producing recombinant CHO cells. J Biotechnol 1999;69:215–226.
17 Borth N, Strutzenberger K, Kunert R, Steinfellner W, Katinger H: Analysis of changes during subclone development and ageing of human antibody-producing heterohybridoma cells by Northern blot and flow cytometry. J Biotechnol 1999;67:57–66.
18 Kromenaker SJ, Srienc F: Stability of producer hybridoma cell lines after sorting: A case study. Biotechnol Prog 1994;10:299–307.
19 Couture ML, Heath CA: Relationship between loss of heavy chains and the appearance of nonproducing hybridomas. Biotechnol Bioeng 1995;47:270–275.
20 Petch D, Butler M: The effect of alternative carbohydrates on the growth and antibody production of a murine hybridoma. Appl Biochem Biotechnol 1996;59:93–104.
21 Miller WM, Blanch HW, Wilke CR: A kinetic analysis of hybridoma growth and metabolism in batch and continuous suspension culture: Effect of nutrient concentration, dilution rate, and pH. Biotechnol Bioeng 2000;67:853–871.
22 Lee MS, Lee GM: Hyperosmotic pressure enhances immunoglobulin transcription rates and secretion rates of KR12H-2 transfectoma. Biotechnol Bioeng 2000;68:260–268.
23 DeZengotita VM, Schmelzer AE, Miller WM: Characterization of hybridoma cell responses to elevated pCO(2) and osmolality: Intracellular pH, cell size, apoptosis, and metabolism. Biotechnol Bioeng 2002;77:369–380.
24 Mocikat R, Kardinal C, Selmayr M: Genetic stability of gene-targeted immunoglobulin loci. 2. Influence of the cell line and the vector linearization site. Mol Gen Genet 1997;256:499–508.
25 Mocikat R: Improving the expression of chimaeric antibodies following homologous recombination in hybridoma cells. J Immunol Methods 1999;225:185–189.

Original Paper

Medical Principles and Practice

Med Princ Pract 2005;14(suppl 1):92–97
DOI: 10.1159/000086190

Received: October 20, 2004
Revised: March 21, 2005

Very-Long-Chain Fatty Acids Activate Lysosomal Hydrolases in Neonatal Human Skin Tissue

Gursev S. Dhaunsi[a] Mazen Al-Essa[a] Wisam Muawad[a] Braham S. Srivastava[a] Nabil Rashwan[b]

[a]Department of Pediatrics, Faculty of Medicine, Kuwait University, [b]Maternity Hospital, Kuwait

Key Words
Peroxisomes · Lysosomes · Catalase · Hydrolases · Very-long-chain fatty acid

Abstract
Objective: The aim of this study was to examine the in vitro effect of peroxisomal dysfunction on lysosomal enzymes, the autophagic machinery in the cell, in order to understand the mechanisms of pathogenesis of peroxisomal disorders. ***Materials and Methods:*** Foreskin samples were obtained immediately after circumcision of 1- to 2-day-old infants at the Maternity Hospital, Kuwait. Skin tissues were cleaned, cut into slices of 1–2 mm^2 in size and treated with lignoceric acid (1–20 μg/ml), a very-long-chain fatty acid (VLCFA), in the presence or absence of 1–5 m*M* aminotriazole (ATZ). A battery of lysosomal enzymes were assayed following treatment of dermal tissue with VLCFA or ATZ. ***Results:*** Treatment of skin slices with lignoceric acid significantly increased ($p < 0.001$) the enzymic activities of acid lipase, acid phosphatase, α-glucosidase, α-galactosidase, N-acetyl-α-*D*-glucosaminidase (NAGA) and N-acetyl-α-*D*-galactosaminidase (NAGTA). ATZ (1–5 m*M*), an inhibitor of key peroxisomal enzyme catalase, also markedly increased the enzymic activities of acid phosphatase, α-glucosidase (23%) and α-galactosidase (18%) without any significant effect on NAGA or NAGTA. Western blot analysis further revealed that both VLCFA and ATZ significantly increased the protein expression of lysosomal enzymes, β-galactosidase and β-glucuronidase. ***Conclusion:*** Experimental dysfunction of peroxisomes mimicked by elevated VLCFA or ATZ-mediated catalase inhibition significantly increased the activities of lysosomal hydrolases in human dermal tissue, suggesting that activation of the lysosomal system could be one of the factors responsible for cellular damage during pathogenesis of peroxisomal diseases.

Introduction

Peroxisomes are known to carry out a number of important cellular functions such as synthesis of plasmalogens, β-oxidation of very-long-chain fatty acids (VLCFAs) and detoxification of reactive oxygen species [1–3]. The importance of peroxisomes in human health was highlighted with the recognition of a number of inherited metabolic disorders associated with dysfunction of this subcellular organelle [4, 5]. Accumulation of VLCFAs, lack of plasmalogens and deficient activities of peroxisomal

KARGER

Fax +41 61 306 12 34
E-Mail karger@karger.ch
www.karger.com

1011–7571/05/0147–0092$22.00/0

Accessible online at:
www.karger.com/mpp

Dr. Gursev S. Dhaunsi
Department of Pediatrics, Faculty of Medicine, Kuwait University
PO Box 24923, 13110 Safat (Kuwait)
Tel. +965 531 9486, Fax +965 533 8940
E-Mail dhaunsig@hotmail.com or dhaunsig@hsc.edu.kw

enzymes are well-documented features of peroxisomal biogenesis disorders [6, 7]. Most of the peroxisomal disorders including the single enzyme defect, X-linked adrenoleukodystrophy are neurodegenerative [8, 9]; however, the pathogenic mechanisms associated with this group of inherited diseases are not clear. Lysosomal hydrolases, which play an important role in various pathophysiological events, have recently been shown to participate in the neurodegenerative process in Alzheimer's disease [10, 11]. The lysosomal system participates in cellular catabolism either by digesting the exogenous material taken up by endocytosis or by breaking down the endogenous material separated from other intracellular components through autophagy. Lysosomal hydrolases are also known to be excreted by exocytosis to degrade extracellular matrix components such as glycosaminoglycans and proteoglycans [12]. Extracellular matrix is vital for structural and functional integration of cells in various tissues, and its degradation through activation of lysosomal enzymes [13, 14], such as α-glucosidase, α-galactosidase, N-acetyl-α-*D*-glucosaminidase (NAGA) and N-acetyl-α-*D*-galactosaminidase (NAGTA), may be responsible for tissue damage during pathological states. This study was carried out to examine the effect of lignoceric acid, a VLCFA, on lysosomal hydrolases in order to understand the mechanisms of cellular and tissue degeneration associated with peroxisomal diseases.

Materials and Methods

Materials

Bovine serum albumin, penicillin/streptomycin, fetal bovine serum and lignoceric acid were purchased from Sigma Chemical Company (St. Louis, Mo., USA). Dulbecco's minimum essential medium (DMEM):Ham's F-12 (1:1) and trypsin-EDTA were from GIBCO (Grand Island, N.Y., USA). Substrates used in lysosomal enzyme assays were obtained from Sigma. Monoclonal antibodies raised against lysosomal glucosidase and glucuronidase proteins were used in this study (MONOSAN, Uden, The Netherlands). Electrophoresis reagents were purchased from Bio-Rad (Richmond, Calif., USA). All other chemicals and reagents were of highest quality and obtained from Sigma and Calbiochem (La Jolla, Calif., USA).

Methods

Preparation of Dermal Tissue and Treatment with Experimental Agents. Foreskin tissues from 1- to 2-day-old babies were obtained immediately after circumcision at a local maternity hospital and processed for experiments. Dermal tissue was washed repeatedly with Eagle's MEM supplemented with 5% fetal calf serum, 5 U/ml penicillin G sodium and 5 μg/ml streptomycin, and cut into small slices. Dermal tissue slices were then incubated at 37°C for 24 h in DMEM F-12 in the presence or absence of varying concentrations (1–5 m*M*) of aminotriazole (ATZ; inhibitor of catalase) and/or VLCFA (1–20 μg/ml of lignoceric acid). After each incubation, the medium was removed and stored at –20°C, and tissue was homogenized in 0.25 *M* sucrose (pH 7.4) buffer.

Assay of Lysosomal Enzymes and Catalase. Enzyme activities of lysosomal hydrolases, acid phosphatase, acid lipase, α-glucosidase, α-galactosidase, NAGA and NAGTA were assayed in the tissue homogenates using 4-methyl-umbelliferone (4-MeU)-linked or *p*-nitrophenol-linked substrates as described elsewhere [16]. Briefly, enzyme reaction was started by adding a known amount of tissue protein to the reaction mixture that consisted of 50 m*M* sodium acetate buffer, pH 4.5, 0.05% Triton X-100 and 0.05 m*M* substrate (4-MeU-α-glucoside or 4-MeU-α-galactoside or *p*-nitrophenyl-NAGA or *p*-nitrophenyl-NAGTA). Enzyme reaction was carried out at 37°C for 15–30 min and then stopped by adding either 0.2 *M* glycine buffer, pH 10.5, or 0.1 *N* sodium hydroxide to the reaction mixture tube. The release of 4-MeU or *p*-nitrophenol following substrate hydrolysis by the lysosomal enzyme was measured using Hoefer's TKO 100 fluorometer or GBC spectrophotometer, respectively. Catalase, a key peroxisomal enzyme, was assayed by the method of Baudhuin et al. [15] using hydrogen peroxide as enzyme substrate.

Western Blot Analysis. Tissue homogenates were prepared following each experiment and SDS-polyacrylamide gel electrophoresis was performed using Laemelli's sample buffer. Equal amounts of proteins for each sample were loaded onto the gel. Following SDS-polyacrylamide gel electrophoresis, proteins were transferred onto nitrocellulose and immunoblotted with antibodies against different lysosomal enzymes. Protein bands were visualized using horseradish peroxidase-labeled secondary antibody and ECL reagents.

Data Analysis

Specific activities of lysosomal hydrolases were calculated for each experiment and data were analyzed for statistical significance using Student's t test.

Results

ATZ (1–5 m*M*) significantly inhibited ($p < 0.001$) the enzymic activity of catalase, resulting in a significant increase ($p < 0.01$) in the enzymic activity of acid phosphatase, a key lysosomal marker enzyme as shown in figure 1. Lignoceric acid, a biochemical characteristic of peroxisomal dysfunction, also significantly ($p < 0.01$) increased the enzymic activities of lysosomal hydrolases, acid phosphatase and acid lipase as shown in figure 2. Activities of lysosomal enzymes involved in degradation of complex glycoconjugate molecules were also assayed, and figure 3 depicts that the enzymic activity of α-glucosidase was significantly ($p < 0.01$) increased following treatment with ATZ and/or lignoceric acid. Enzymic activity of α-galactosidase (fig. 4), another hydrolytic enzyme for complex glycosaminoglycan molecules, was also markedly elevated ($p < 0.01$) by experimental conditions of peroxisomal

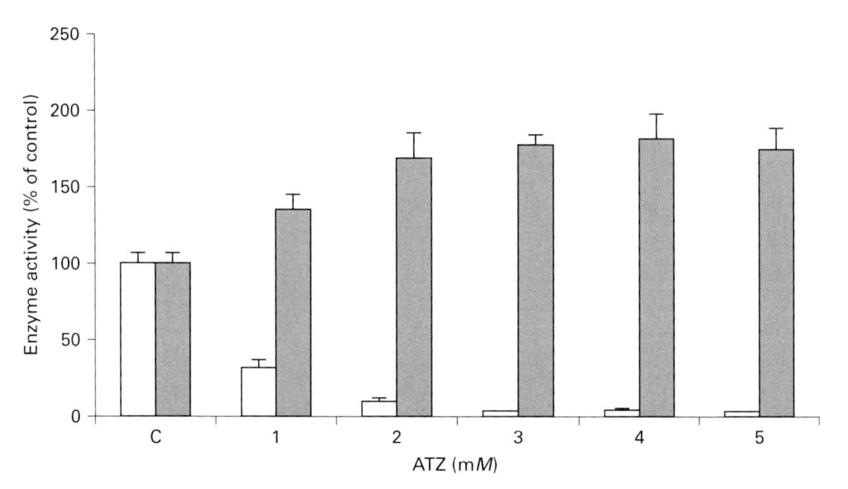

Fig. 1. Enzyme activities of catalase (open bars) and acid phosphatase (dark bars) in human dermal tissue following treatment with 1–5 mM of ATZ. Results shown are means ± SD of specific enzyme activities (percent of control) from five different experiments. Control specific activity of acid phosphatase is 14.28 ± 0.81 μmol/h/mg protein.

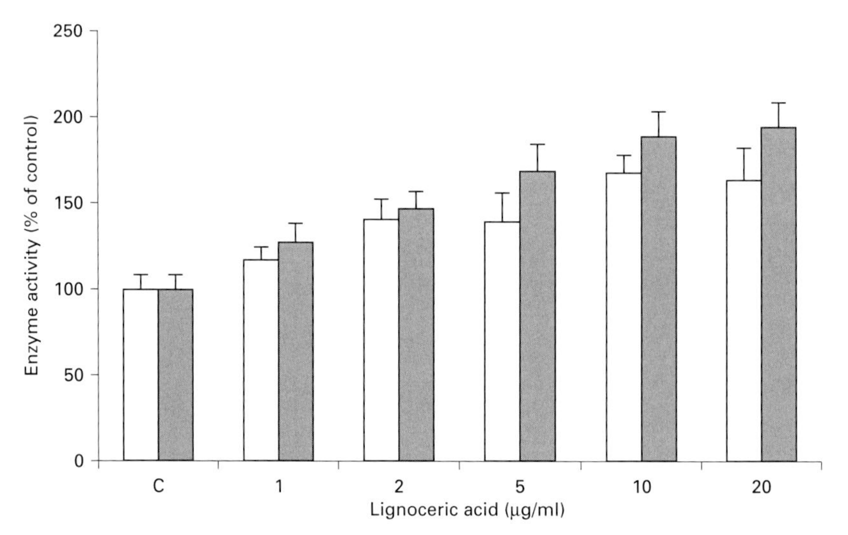

Fig. 2. Specific enzyme activities (as percent of control) of acid lipase (open bars) and acid phosphatase (dark bars) in human dermal tissue following treatment with lignoceric acid (1–20 μg/ml). Values are means ± SD of five different experiments. Control specific activity of acid lipase is 4.21 ± 0.26 μmol/h/mg protein.

dysfunction; however, ATZ and VLCFA did not synergize each other's effect on enzymic activities of lysosomal hydrolases. Figures 5 and 6 show that enzymic activities of NAGA and NAGTA were not influenced by inhibition of catalase, whereas lignoceric acid significantly ($p < 0.01$) enhanced the activities of both lysosomal hydrolases, NAGA and NAGTA. Though VLCFA and/or ATZ enhanced the activities of various lysosomal enzymes, none of these lysosomal proteins leaked out of the tissues as no activity could be detected for any of these lysosomal hydrolases in the incubation media. In order to assess whether the observed effects of catalase inhibition and/or lignoceric acid in this study had a transcriptional and/or translational regulation, Western blot analysis was performed, and figure 7 shows that both ATZ and lignoceric acid significantly (2- to 3-fold) increased the protein expression of lysosomal hydrolases, β-galactosidase and β-glucuronidase. Overall, lignoceric acid was observed to be a versatile and highly potent modulator of lysosomal hydrolases, whereas inhibition of catalase resulted in selective activation of lysosomal enzymes. Further, the lack of synergism between the effects of VLCFA and catalase inhibition on lysosomal hydrolases suggests that two different mechanisms might be involved in the observed activation of the lysosomal hydrolytic machinery during the pathogenesis of peroxisomal disorders.

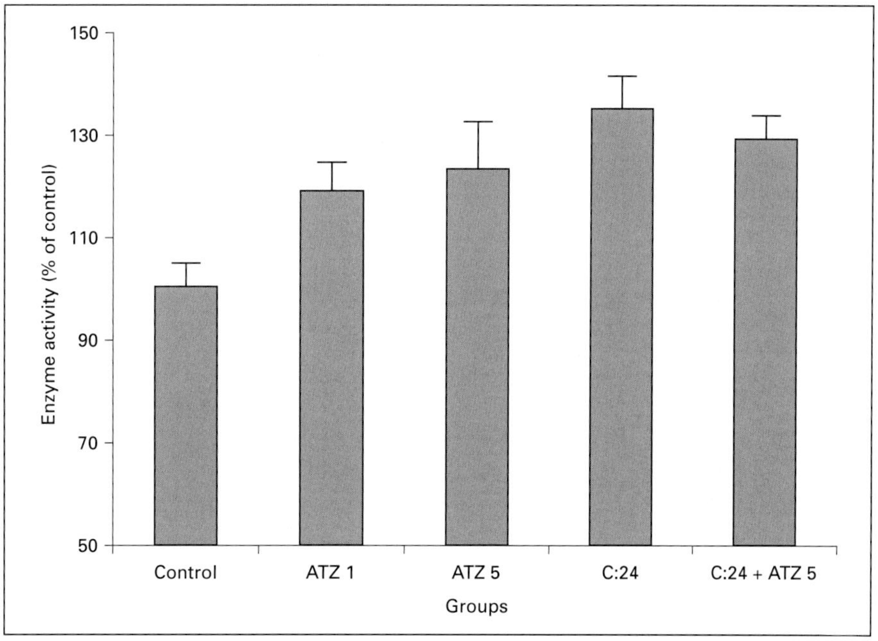

Fig. 3. Enzyme activity (as percent of control) of α-glucosidase in human dermal tissue following treatment with ATZ, 1 m*M* (ATZ 1) or 5 m*M* (ATZ 5) in the presence or absence of lignoceric acid (C:24). Results shown are means ± SD of specific enzyme activities (as percent of control) from five different experiments. Control represents tissue without ATZ or C:24 treatment. Control specific activity of α-glucosidase is 9.41 ± 0.18 μmol/h/mg protein.

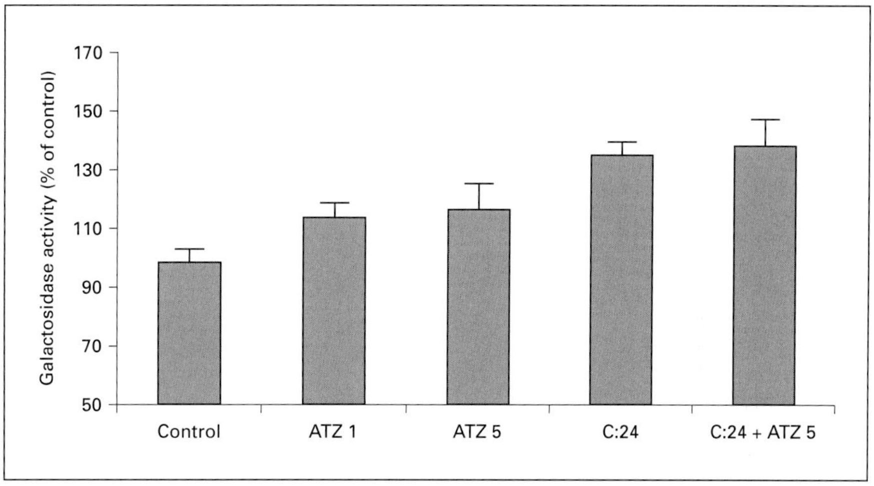

Fig. 4. α-Galactosidase activity in human dermal tissue treated with ATZ, 1 m*M* (ATZ 1) or 5 m*M* (ATZ 5) in the presence or absence of lignoceric acid (C:24). Control specific activity of α-galactosidase is 12.4 ± 1.8 μmol/h/mg protein. Results shown are means ± SD of specific enzyme activities (as percent of control) from five different experiments.

Discussion

Peroxisomal disorders are a relatively new category of autosomal inherited metabolic diseases, but represent a deadly class of genetic disorders that are commonly associated with developmental delay, hypotonia, failure to thrive and neurological impairment [4, 17]. Due to a defect in the assembly of the organelle or import of proteins into the organelle, peroxisomal functions are impaired in patients with peroxisomal diseases. Elevated levels of VLCFAs due to impaired β-oxidation, and increased oxidative stress due to deficiencies of peroxisomal enzymes like catalase are believed to cause cellular pathology in peroxisomal disorders [18, 19]. Though the biochemical defects linked with peroxisomal disorders have been identified, the molecular mechanisms of pathogenesis that lead to cellular and tissue degeneration are unknown. Upregulation of lysosomes, the subcellular organelles that carry out degradation of several types of complex biomolecules, has recently been suggested to play a role in the pathogenesis of neurodegenerative disorders like Alzheimer's disease [10, 20]. Activation of lysosomal hydrolases, such as acid phosphatase, acid lipase, α-glucosidase and α-galactosidase, by ATZ or lignoceric acid as observed in this study suggests that the lysosomal system may be involved in the pathogenesis of peroxisomal dis-

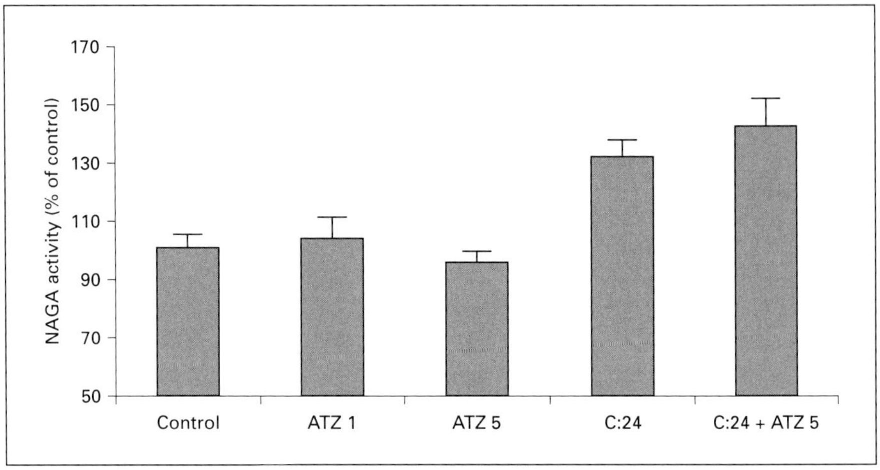

Fig. 5. Effect of ATZ, 1 m*M* (ATZ 1) or 5 m*M* (ATZ 5) on enzyme activity of NAGA with or without lignoceric acid (C:24) treatment. Control values for NAGA are 5.48 ± 0.21 μmol/h/mg protein. Results shown are means ± SD of specific enzyme activities (as percent of control) from five different experiments.

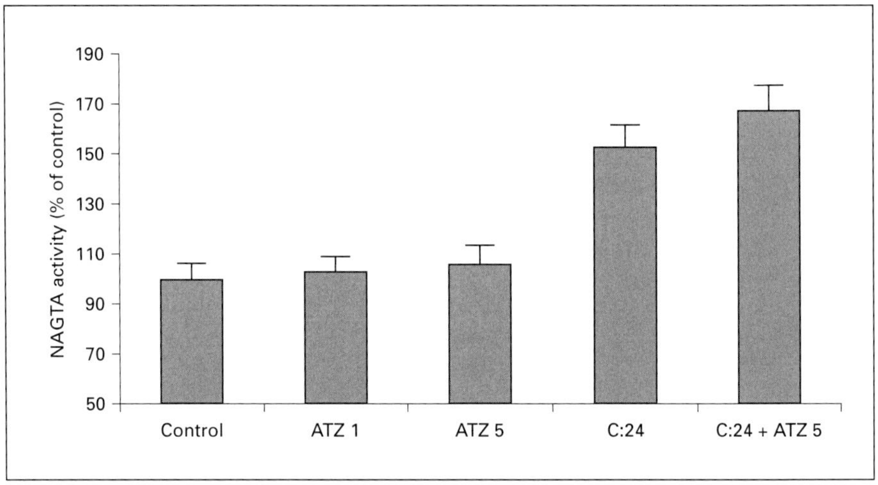

Fig. 6. NAGTA activity in human dermal tissue following treatment with ATZ, 1 m*M* (ATZ 1) or 5 m*M* (ATZ 5), and/or lignoceric acid (C:24). Control NAGTA activity is 4.62 ± 0.33 μmol/h/mg protein. Results shown are means ± SD of specific enzyme activities (as percent of control) from five different experiments.

eases as well. Stimulating effects of VLCFA and/or ATZ on lysosomal enzymes as observed in this study are intriguing in terms of molecular mechanisms of the pathogenesis of peroxisomal diseases. Reactive oxygen species are known to affect several cellular functions and activation of lysosomal enzymes by hydrogen peroxide or superoxide anions is documented to contribute towards cellular pathology [21, 22]. Inhibition of catalase by ATZ would be expected to raise the intracellular levels of hydrogen peroxide, thus, the observed activation of lysosomal enzymes by ATZ in this study is most likely due to increased cellular oxidative stress. Though free fatty acids or fatty acid esters are believed to affect the lysosomal system either through signal transduction mechanisms involving PKC or by destabilizing the lysosomal membranes [23], the effect of VLCFAs on lysosomal enzymes has remained unexplored. This study shows for the first time that VLCFAs lead to activation of the lysosomal

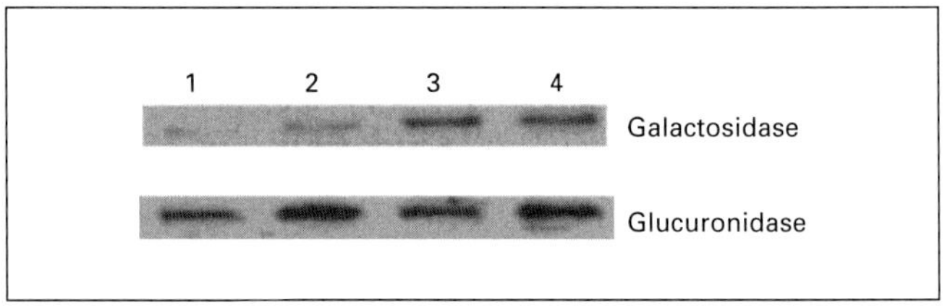

Fig. 7. Immunoblot detection of lysosomal enzymes, β-galactosidase and β-glucuronidase, in human dermal tissue treated with lignoceric acid (lane 2) or 5 m*M* ATZ (lane 3) or both (lane 4). Control is shown in lane 1.

system, thereby suggesting another key mechanism for the pathogenesis of peroxisomal disorders. A selective and nonsynergistic increase in lysosomal enzyme activities by ATZ and lignoceric acid as observed in this study indicates that VLCFAs and oxidative stress may be up-regulating the lysosomal system through different and independent mechanisms. Both ATZ and lignoceric acid increase the enzyme activities of acid phosphatase, α-galactosidase and α-glucosidase; however, the activation of NAGA and NAGTA is only specific to lignoceric acid. Pathological conditions associated with degradation of connective tissue or extracellular matrix have earlier been shown to exhibit increased activities of lysosomal enzymes [24, 25]. In this in vitro study, the lignoceric acid-mediated increase in activities of NAGA and NAGTA, two lysosomal hydrolases mainly responsible for the breakdown of glycosaminoglycans of the extracellular matrix, would further suggest a possible degradation of intercellular connective tissue during peroxisomal dysfunction in vivo. Our findings that lignoceric acid significantly enhances the activities of lysosomal enzymes, which carry out degradation of extracellular matrix components and other complex intracellular molecules, further underscores the role of lysosomal hydrolases in the pathogenesis of disorders of peroxisomal fatty acid oxidation such as X-linked adrenoleukodystrophy. Future studies should, however, reveal if the activities of lysosomal hydrolases are elevated in the tissues of peroxisomal disorder patients and whether such induction of the lysosomal system is at the transcriptional level.

Conclusions

An increased activity of lysosomal hydrolases by the presence of elevated VLCFAs is an important finding with regard to understanding the cellular and molecular mechanisms of the pathogenesis of peroxisomal dysfunction. Activation of the lysosomal system by the events of peroxisomal defect may destabilize tissue organization through increased autophagy-, endocytosis- or exocytosis-mediated degradation of complex biomolecules and cause tissue damage such as neurodegeneration.

References

1 Hajra AK, Bishop JE: Glycerolipid biosynthesis in peroxisomes via the acyl dihydroxyacetone phosphate pathway. Ann NY Acad Sci 1982;386:170–182.
2 Lazarow PB, de Duve C: Fatty acyl-CoA oxidizing system in rat liver peroxisomes, enhancement by clofibrate, a hypolipidemic drug. Proc Natl Acad Sci USA 1976;73:2043–2046.
3 Dhaunsi GS, Gulati S, Singh AK, Orak JK, Asayama K, Singh I: Demonstration of Cu-Zn superoxide dismutase in rat liver peroxisomes: Biochemical and immunochemical evidence. J Biol Chem 1993;267:6870–6873.
4 Lazarow PB, Moser HW: Disorders of peroxisome biogenesis; in Scriver CR, Beaudet AL, Sly WS, Valle D (eds): The Metabolic and Molecular Bases of Inherited Disease. New York, McGraw-Hill, 1995, pp 2287–2324.
5 Brown FR, Voigt R, Singh AK, Singh I: Peroxisomal disorders: Neurodevelopmental and biochemical aspects. Am J Dis Child 1993; 147:617–626.
6 Wanders RJ, Tager JM: Lipid metabolism in relation to human disease. Mol Aspects Med 1998;19:69–154.
7 Poggi-Travert F, Fournier B, Poll-The BT, Saudubray JM: Clinical approach to inherited peroxisomal disorders. J Inherit Metab Dis 1995;18(suppl):1–18.
8 Schutgens RBH, Heymans HAS, Wanders RJA, Van den Bosch H, Tager JM: Peroxisomal disorders: A newly recognized group of genetic diseases. Eur J Pediatr 1986;144:430–440.
9 Wanders RJ, Heymans HS, Schutgens RB, Barth PG, Van den Bosch H, Tager JM: Peroxisomal disorders in neurology. J Neurol Sci 1988;88:1–39.
10 Kobayashi K, Hayashi M, Nakano H, Fukutani Y, Sasaki K, Shimazaki M, Koshino Y: Apoptosis of astrocytes with enhanced lysosomal activity and oligodendrocytes in white matter lesions in Alzheimer's disease. Neuropathol Appl Neurobiol 2002;28:238–251.
11 Mathews PM, Guerra CB, Jiang Y, Grbovic OM, Kao BH, Schmidt SD, Dinakar R, Mercken M, Hille-Rehfeld A, Rohrer J, Mehta P, Cataldo AM, Nixon RA: Alzheimer's disease-related expression of the cation-dependent mannose 6-phosphate receptor increases Abeta secretion: Role for altered lysosomal hydrolase distribution in beta-amyloidogenesis. J Biol Chem 2002;277:5299–5307.
12 Pennock CA: Lysosomal storage disorders; in Holton JB (ed): The Inherited Metabolic Diseases, ed 2. New York, Churchill Livingstone, 1994, pp 205–242.
13 Barret AJ, Heath MF: Lysosomal enzymes; in Dingle JT (ed): Lysosomes: A Laboratory Handbook, ed 2. Amsterdam, North Holland, 1977, pp 19–79.
14 Olczyk K, Kucharz EJ, Sonecki P: Matrix-degrading lysosomal hydrolases in serum of rats with hepatic fibrosis treated with methylthiouracil or thyroideum. Rom J Endocrinol 1993; 31:23026.
15 Baudhuin P, Beaufay Y, Rahman L, Sellinger OZ, Wattiaux R, Jacques P, de Duve C: Tissue fractionation studies. Biochem J 1964;92:179–184.
16 Ozand PT, Gascon G, Al-Aqeel A, Roberts G, Dhalla M, Subramanyam SB: Prevalence of different types of lysosomal storage diseases in Saudi Arabia. J Inherit Metab Dis 1990;13: 849–861.
17 Kamei A, Houdou S, Takashima S, Suzuki Y, Becker LE, Armstrong DL: Peroxisomal disorders in children: Immunohistochemistry and neuropathology. J Pediatr 1993;122:573–579.
18 Moser HW: Peroxisomal disorders. Clin Biochem 1991;24:343–351.
19 Singh I: Peroxisomal fatty acid oxidation and cellular oxidative stress. Methods Enzymol 2002;352:361–372.
20 Adamec E, Mohan PS, Cataldo AM, Vonsattel JP, Nixon RA: Up-regulation of the lysosomal system in experimental models of neuronal injury: Implications for Alzheimer's disease. Neuroscience 2000;100:663–675.
21 Gupta A, Sharma N, Sharma BK, Sharma S, Ganguly NK: Oxygen-dependent and -independent mechanisms of renal injury in experimental ascending pyelonephritis. FEMS Immunol Med Microbiol 1996;13:35–42.
22 Meera R, Anand S, Ramesh CV, Puvanakrishnan R: Inhibition of neutrophil derived lysosomal enzymes and reactive oxygen species by a novel tetrapeptide. Inflamm Res 1999;48: 479–484.
23 Boneh A: A model for PKC involvement in the pathogenesis of inborn errors of metabolism. Trends Mol Med 2002;8:524–531.
24 Dickinson DP: Cysteine peptidases of mammals: Their biological roles and potential effects in the oral cavity and other tissues in health and disease. Crit Rev Oral Biol Med 2002;13:238–275.
25 Wolters PJ, Chapman HA: Importance of lysosomal cysteine proteases in lung disease. Respir Res 2000;1:170–177.

Medical Principles and Practice

Applications of Molecular Biology in Modern Medicine

Med Princ Pract 2005;14(suppl 1):98–109
DOI: 10.1159/000086191

Abstracts

Hematology

1
Molecular Genetics of Blood Disorders: A Basis for New Approaches to Management

L. Luzzatto

National Institute for Cancer Research, Genova, Italy

Understanding the nature of disease has been a major aim of medicine for centuries in all cultures. In contemporary medicine, this effort has continued, and generally speaking, we can regard the aim as successfully achieved when the molecular basis of a disease has been elucidated. It can be safely stated that haematology has often been at the forefront in this process, starting from the historical discovery, now half a century old, that sickle cell anaemia is due to a minute change in the haemoglobin molecule. As a result of this discovery, the phrase *Molecular Disease* was coined. Today the molecular genetics of inherited blood disorders is well advanced. In the case of *red cells,* the majority of these manifest as haemolytic anaemias: they include haemoglobin disorders, enzyme disorders and membrane disorders. The underlying molecular lesions have been identified, approximately in that time sequence. In terms of health burden, haemoglobinopathies, thalassaemias and G6PD deficiency, taken together, constitute probably the set of Mendelian diseases that has the greatest impact on human populations. In spite of the wealth of genetic information on blood groups, molecular analysis of these systems has lagged behind for some time: but now the majority of blood group genes have been cloned, and their structure has largely explained, for instance the bewildering complexity of the rhesus system. Inherited disorders of *white cells* and *platelets* are much less prevalent: however, elucidating the molecular genetics of chronic granulomatous disease has helped to understand not only this condition, but also the physiology of the oxidative burst, a critically important feature in the role of neutrophils in defending us against bacteria. Disorders of *coagulation* have probably been amongst the first for which the inherited nature has been appreciated: today, the molecular genetics of the entire coagulation cascade is essentially solved. Equally important, we know the genes for many of the coagulation inhibitors that keep the process under a finely tuned control. In the area of acquired disorders, once again, we have had landmarks of historic significance in haematology. A unique type of haemolytic anaemia, paroxysmal nocturnal haemoglobinuria, was shown in 1970 to be a clonal disorder. At about the same time, the study of chronic myeloid leukaemia and of Burkitt's lymphoma linked for the first time specific chromosomal translocations to individual forms of malignancy. In the early eighties, the respective breakpoints were characterized at the molecular level: *bcr-abl* became the prototype of a large number of fusion genes, while *IgH-myc* heralded the notion that a tumour could arise from dysregulation of a proto-oncogene through its juxtaposition to a gene that is highly expressed in a particular cell lineage. Thus, in terms of understanding blood disorders in depth, the molecular approach has been truly revolutionary: fortunately, it has already paid dividends in terms of clinical implications as well. First of all, molecular analysis can help definitive diagnosis, and sometimes it is the only way to identify subjects who are at high risk, as in the case of haemochromatosis. Second, molecular diagnosis of an inherited disorder can reliably identify heterozygotes and can be carried out even on cells in which the relevant gene is not expressed. Thus, testing amniotic fluid cells or chorionic villi makes prenatal diagnosis possible. Screening for heterozygotes combined with prenatal testing can be an effective approach to prevention of serious inherited disorders. In fact, this approach has been the foundation for several systematic programmes that have had a significant impact on the frequency of inherited disorders: perhaps the most spectacular example has been the decrease in the incidence of β-thalassaemia major in Cyprus and in Sardinia. Finally, the ultimate goal of medicine is to cure patients who are ill, whatever the original cause or mechanism. In the past, certain diseases have been regarded as incurable because they are inherited: however, the precise knowledge of their molecular lesions may be used to reverse the argument. Indeed, since the inception of recombinant DNA technology, the possibility to correct inherited abnormalities by gene transfer has been keenly entertained. Until recently, this may have seemed an overambitious goal: however, over the past 5 years, several inherited conditions have been successfully treated by gene transfer into haematopoietic stem cells in preclinical models in animals. Most important, in several children, the clinical manifestations of two types of severe combined immunodeficiency have been reversed, perhaps permanently, by retroviral-mediated gene transfer of adenosine deaminase and of the cytokine γ receptor, respectively. Thus, gene therapy has at long last become a reality.

KARGER

Fax +41 61 306 12 34
E-Mail karger@karger.ch
www.karger.com

1011–7571/05/0147–0098$22.00/0

Accessible online at:
www.karger.com/mpp

2

Molecular Studies on Genetic Risk Factors for Venous Thrombo-Embolism in Kuwait

K.C. Das, M.M. Jadaon, M.M. Al-Otaibi, R. Marouf, S. Al-Humood, K. Easow Sicy, H. Al-Wazan, A. Gupta, A. Khamis, A.R. Samia

Haematology Unit, Department of Pathology, Faculty of Medicine, Kuwait University; Haematology Unit, Mubarak Al-Kabeer Hospital and Amiri Hospital, Kuwait

Venous thrombo-embolism (VTE) is a common cardiovascular disorder with a potential for serious consequences. This occurs frequently in Kuwait and is more common in women than in men. VTE appears to affect a relatively large proportion of young people before the age of 40 years, often without an apparent precipitating cause. These circumstantial events elicit the possibility of genetic predisposition as an important factor in its genesis. Several genetic disorders involving the haemostatic process have been described in patients with VTE. Among these genetic risk factors, two mutations, factor V Leiden (FVL) and prothrombin G20210A, have been reported to be prevalent among patients of Caucasian origin, and rare or absent in the Chinese, Japanese and other South-East Asian and African patients with VTE. The present work was undertaken to determine the presence of genetic risk factors with particular reference to molecular genetic studies on FVL and prothrombin G20210A mutations in 1,075 ethnic Arab patients with VTE in Kuwait. Natural inhibitors of blood coagulation in the patients' plasma were estimated by chromogenic and clotting (functional) assays. Molecular analysis of DNA for FVL and prothrombin G20210A mutations were performed by PCR amplification of patients' DNA using carefully selected sets of primers, followed by appropriate restriction enzyme treatment and agarose/polyacrylamide gel electrophoresis. The presence of mutations was further confirmed by DNA sequencing using Sanger's dideoxynucleotide chain termination method in an automated DNA sequencer. Analysis of the results shows that approximately a little over 50% of the patients with VTE had one or more genetic risk factors. Among them, about 16% had an FVL mutation, 6% had the prothrombin G20210A mutation, 10% had protein C deficiency, 14% had protein S deficiency and 4.5% had AT-III deficiency. The presence of the prothrombin G20210A mutation was consistently associated with markedly increased plasma levels of prothrombin activity. A small proportion of patients with FVL mutation also had prothrombin G20210 mutation, or other genetic risk factors such as protein C deficiency, protein S deficiency or AT-III deficiency. The concomitant presence of more than one genetic risk factors appears to cause an additive effect on the risk of development of VTE and augment the chances of their recurrence. Thus, the prevalence of genetic risk factors for VTE including FVL and prothrombin G20210A mutations is high in Kuwait, where Europe, Asia and Africa meet geographically and is comparable to that in patients of Caucasian origin. These observations make a good case for considering the introduction of a genetic screening program for thrombotic disorders in the health care system.

Supported by grants from the College of Graduate Studies, Kuwait University, Kuwait.

3

Anthropological and Clinical Implications of Sickle Cell Haplotypes in the Arabian Peninsula

A. Adekile

Department of Pediatrics, Faculty of Medicine, Kuwait University, Kuwait

Reports of Peninsular Arab sickle cell anemia patients, surviving into late adulthood, with little or no clinical problems, started appearing in the literature in the late 1960s and early 1970s. This was contrary to experience elsewhere and was the beginning of the notion that indeed, the disease has considerable clinical heterogeneity. The onset of DNA recombinant technology in the 1970s afforded extensive investigation of this genotype/phenotype discordance. Several polymorphic restriction endonuclease sites were identified in the β-globin gene cluster and five common haplotypes emerged. These were designated Benin (BEN), Senegal (SEN) Bantu (BAN), Saudi Arabia/India (SAI) and Cameroon (CAM), named for the area where each was most prevalent. Further studies led to the conclusion that the β^S mutation (GAG→GTG) must have occurred independently in each of these five locations and subsequently spread by migration. It is also recognized that the phenotypic expression of the disease is determined to a large extent by the haplotype. Thus, the SEN and SAI are associated with the mildest, while BAN is associated with the most severe disease. The mutation on the SAI haplotype probably originated in the Indus Valley and was dispersed during the Harappan culture. However, there had been long-standing interaction between the prehistoric peoples of the East African coast and those in Yemen and Oman. All these anthropological antecedents influence the present-day distribution of β^S haplotypes in the Arabian Peninsula. Thus, while BEN predominates in Western Saudi Arabia and Yemen, SAI predominates in Eastern Saudi Arabia, Kuwait, Bahrain and United Arab Emirates; in Oman, there is a coexistence of three major haplotypes (52.1% BEN, 26.7% SAI and 21.4% BAN). Although the symptomatology and complications of sickle cell disease is well documented in patients with other haplotype backgrounds, the features of the disease among Arab patients with the SAI haplotype have been systematically studied only recently. Indeed these patients have a milder phenotype, probably because of their persistently high fetal hemoglobin (Hb) levels (~20%) consequent on the Xmn1 positive site in the promoter region of the $^G\gamma$-globin gene. They have a higher Hb level, with a reduced degree of hemolysis and are protected from many of the complications of the disease, e.g. stroke, priapism, chronic leg ulcers and severe bacterial infections because of preserved spleen function. However, the patients may be more predisposed to osteonecrosis from early childhood. A recent study from Kuwait found that 26% of patients between the ages of 6 and 16 years have evidence of avascular necrosis of the femoral head on magnetic resonance imaging studies. It is becoming clear that even among these patients with high Hb F levels, the phenotype is not uniform; some patients have more frequent problems than others. The modulating factors in this group need further elucidation.

4

Clinical Significance of Trisomy 5 in Acute Lymphoblastic Leukemia

M. Alkhalaf [a], R. Marouf [b], S. Al-Humood [b], M.H.A. Burhamah [c]

Departments of [a]Biochemistry and [b]Pathology, Faculty of Medicine, Kuwait University, and [c]Pediatric Hematology Unit, Al-Sabah and NBK Pediatric Hospital, Kuwait

Trisomy 5 is a rare event in acute lymphoblastic leukemia (ALL), except in the context of a hyperdiploid leukemic clone. Due to the limited number of cases identified so far, the clinical significance of trisomy 5 as the sole cytogenetic aberration in ALL is not fully appreciated. The aim of this study was to report a new case of trisomy 5 as the sole cytogenetic abnormality in infantile precursor B ALL. The patient is the youngest (22-month-old boy) to be reported with trisomy 5 as the only karyotypic anomaly in ALL. The cytogenetic aberration was identified at the time of diagnosis and before initiation of treatment. Nine months after initiation of chemotherapy, the patient remains in complete clinical and cytogenetic remission. The clinical and cytogenetic findings in this patient are presented in relation to other known cases of ALL with trisomy 5. To the authors' knowledge, this is the first case of trisomy 5 as the sole cytogenetic anomaly in ALL to be reported at this early age. The observed trisomy 5 may be responsible for the rapid and favorable response to chemotherapy in this patient.

5

Molecular Analysis of A^1, A^2, B, O^1 and O^2 Alleles by Multiplexed PCR-RFLP at the ABO Locus in a Kuwaiti Population

M.M. El-Zawahri [a], Y.A. Luqmani [b], G. Abraham [b], A. Al-Bashir [c]

[a]Department of Biological Sciences, Faculty of Science and [b]Department of Pharmaceutical Chemistry, Faculty of Pharmacy, Kuwait University, and [c]Central Blood Bank, Kuwait

The ABO blood group system has been one of the most widely studied genetic polymorphisms and has long been in use in forensic serology, blood transfusion, paternity testing and anthropologic investigation. Genomic DNA was extracted from blood samples taken from 115 Kuwaiti donors. The phenotypes of the samples, determined serologically, were A (n = 30), B (n = 32), AB (n = 2) and O (n = 51). With multiplex PCR, fragments of 252 (251 for O^1) and 843 (842 for A^2) bp were amplified in the same tube using primers mo-46/57 and mo-101/71, respectively [Olsson and Chester, 1995]. The amplified fragments cover the two major exons 6 and 7, and constitute 91% of the translated sequence. After simultaneous digestion with restriction endonucleases *Hpa*II and *Kpn*I, a conclusive interpretation of the expected fragment pattern obtained could be made in all cases. Fifteen different genotypes could be identified when combining the A^1, A^2, B, O^1 and O^2 alleles from the digestion patterns. The genotyping results of the 115 samples were : 1 A^1A^1 (0.87%), 2 A^1A^2 (1.74%), 13 A^1O^1 (11.3%), 14 A^2O^1 (12.17%), 2 A^1B (1.74%), 5 BB (4.35%), 27 BO^1 (23.48%), 50 O^1O^1 (43.48%) and 1 O^1O^2 (0.87%). Six of the possible combinations (A^1O^2, A^2A^2, A^2O^2, A^2B, BO^2 and O^2O^2) were not found among the 115 samples tested. All genotyping results were in agreement with the serologically determined phenotypes. The present study is the first demonstration of variability linked to A^1, A^2, B, O^1 and O^2 alleles of the ABO locus in a Kuwaiti population. These genotypic frequency data may provide useful additional information for forensic, population genetics, phylogenetics and evolution studies in this population.

Supported by Kuwait University Research Administration grant SZ 01/00.

Infectious Diseases

6

Signaling Pathways in the Pathogenesis of Viral Hepatitis

S. Jameel

International Centre for Genetic Engineering and Biotechnology (ICGEB), New Delhi, India

Viruses are obligate parasites requiring host functions for their multiplication. Over millions of years of co-evolution of viruses and their hosts, intricate mechanisms have evolved to suit either the virus or the host. The outcome of a viral infection (clearance or persistence) is therefore based largely on how a virus interacts with its host. The interaction of viruses with their hosts is played out over three overlapping responses. These include effects on the host immune response, the cell signaling and cell death (apoptosis) pathways. To clear viral infection, the host uses innate as well as antigen-specific immune responses, including natural killer cells, interferons, antibodies and T cells. Viruses have developed mechanisms to evade one or more of these responses, mediated by virus-encoded proteins. These proteins enable the virus to replicate efficiently in the presence of an antiviral host response. Apoptosis or programmed cell death is a non-specific strategy developed by the host to clear viral (and other) infection. A few cells are sacrificed for the overall benefit to the host. Consequently, viruses have developed strategies to inhibit apoptosis. The immune response and apoptosis are critically dependent upon signal transduction in the host. The development of antibodies and antigen-specific effector T cells require the binding of viral proteins to the surface of memory cells. The resulting signals transduce to the nucleus and modulate the expression of genes critical to development, differentiation and proliferation. Similarly, the killing of virus-infected cells by antigen-specific cytotoxic T cells is critically dependent upon the engagement of and signaling through death receptors. Naturally, in their quest for survival, viruses have evolved strategies to modulate various signaling pathways in the host cell. Five viruses are so far known to cause hepatitis. These include the hepatitis A, B, C and E viruses (HAV, HBV, HCV, HEV) and the delta agent (HDV). Of these, HAV and HEV cause acute self-limited infection, while HBV and HCV cause varying degrees of persistence leading to cirrhosis and hepatocellular carcinoma. Can the clinical outcome of these viral infections be rationalized by understanding their ability to modulate the host immune response, cell death and signaling pathways? Recent studies from many laboratories have characterized proteins encoded by hepatitis viruses that modulate these host pathways to varying degrees. Though all hepatitis viruses are different in terms of their genome organization, gene expression and replication strategies, common themes seem to emerge in how they interact with the host cell. The recognition of such pathways would be critical not just to understanding pathogenesis, but also in developing strategies for the clinical management of these infections.

7
Advances in the Molecular Diagnosis of Infectious Diseases

A. Voevodin

Department of Microbiology, Faculty of Medicine, Kuwait University, Kuwait

Nucleic acid-based tests, commonly referred to as molecular tests, are playing an increasingly important role in the diagnosis of infectious diseases. Molecular tests allow direct detection of microbes in clinical specimens. A number of innovative and proprietary technologies have been introduced in this field. They can be broadly divided into target amplification (examples: polymerase chain reaction, nucleic acid-based amplification, transcription-mediated assay, strand displacement assay) and signal amplification (examples: branched DNA assay, hybrid-specific antibody assay, 'Invader' assay) techniques. According to the signal detection mode, molecular tests can be classified as single end-point and real-time. The former are usually qualitative, the latter are quantitative. Analytical sensitivity of some molecular tests reaches the theoretical maximum, i.e. one genome equivalent per reaction input sample. Analytical specificity of molecular tests is also very high – some of these tests allow nucleic acid targets differing by just one nucleotide to be distinguished. However, this is not always translated into high diagnostic sensitivity and specificity. Molecular tests have many advantages as compared with the classical approaches for direct detection of microbes. They are more rapid, objective, associated with minimal biohazard and can be automated. However, transition of these tests from research to clinical settings is associated with a number of difficulties and restrictions. Some of those are fundamental (high analytical sensitivity and specificity rarely translates into high diagnostic sensitivity and specificity; uncertainty with the 'gold standard'), some are material (high cost) and some are legal (use of all major molecular technologies for diagnostic purposes is restricted by patent laws; very few molecular tests are approved for diagnostic purposes). The major 'development avenues' for molecular diagnostic methodologies are automation, miniaturization and multiplexing (simultaneous detection of multiple microbes). A number of innovative approaches are being pursued in this field. However, skyrocketing technological progress should not mask more 'down-to-earth' issues such as external quality control/assurance (still in its infancy), justified use (there is tendency for 'overuse' of molecular tests) and cost (still is inhibitory for the majority of potential users).

8
Absence of a Specific 530-bp Deletion in the Promoter Region of *Actinobacillus actinomycetemcomitans* Leukotoxin Gene Operon in Strains from Caucasian Adults

H.P. Müller[a, b], A. Macheleidt[c], A. Fuhrmann[b, c], T. Eger[c], L. Zöller[c]

[a]Faculty of Dentistry, Kuwait University, Kuwait; [b]School of Dental Medicine, Heidelberg University, and [c]Central Institute of the German Armed Forces Medical Services, Koblenz, Germany

Introduction and Objectives: Leukotoxin production is an important virulence factor of the periodontal pathogen, *Actinobacillus actinomycetemcomitans.* It belongs to the family of RTX toxins whose target is the β_2-integrin, LFA-1, causing pore formation of cell membranes of neutrophils and monocytes. Strains with a specific 530-bp deletion in the promoter region of the leukotoxin gene operon produce 10–20 times higher amounts of leukotoxin. They have recently been associated with young Afro-Americans, Afro-Indians, individuals from Western Africa, and the Middle East, who were suffering from aggressive periodontitis. In a series of investigations, we attempted to isolate highly leukotoxic strains of *A. actinomycetemcomitans* among young adult Caucasians with plaque-induced gingivitis and patients with chronic or aggressive periodontitis. **Methods:** A total of 298 young adults between 18 and 29 years of age presenting with plaque-induced gingivitis were screened for oral *A. actinomycetemcomitans.* The organism was found in 68 individuals (23%). Primers amplifying a species-specific 16S rRNA sequence, a sequence in the *ltxA* region, and sequences in the leukotoxin promoter region were employed in PCR. In addition, strains from 25 periodontitis patients were analyzed. **Results:** Whereas all strains harbored the leukotoxin gene, the promoter deletion was only found in a strain isolated in a 24-year-old male immigrant from Ghana with severe localized aggressive periodontitis. Overall, no strong association between the presence of minimally leukotoxic *A. actinomycetemcomitans* strains and clinical signs of periodontal disease could be ascertained. **Conclusion:** Highly leukotoxic strains characterized by a 530-bp deletion in the promoter region of the leukotoxin gene seem to be missing in Caucasians.

Supported in part by the Deutsche Forschungsgemeinschaft grant 1404/2-1, Germany.

9
Multiplex PCR for Investigating Population-Based Carriage Patterns of Periodontal Pathogens in Saliva

E. Könönen[a, b], P. Di Tella[a]

[a]Department of Microbiology, National Public Health Institute, Helsinki, Finland; [b]Faculty of Dentistry, Kuwait University, Kuwait

Introduction and Objectives: Multiplex PCR offers a labor-minimizing technique for simultaneous detection of target microbes when a large number of specimens will be processed. We used multiplex PCR to examine the salivary carriage rates of three periodontal pathogens, *Actinobacillus actinomycetemcomitans (Aa), Porphyromonas gingivalis (Pg),* and *Tannerella forsythensis (Tf)* (formerly *Bacteroides forsythus*) in a random sample of 608 adult Helsinki citizens divided into nine age cohorts (30–34, 35–39, 40–44, 45–49, 50–54, 55–59, 60–64, 65–74, and ≥ 75 years). **Methods:** As biological material, saliva contains PCR inhibitors. Therefore, a combination of boiling and cation-chelating resin was used for pretreatment of salivary specimens. Three species-specific 16S rRNA gene forward primers and a conversed universal reverse primer revealed specific amplicons for *Tf, Aa,* and *Pg* at 745, 360, and 197 bp, respectively. **Results:** The protocol detected even 1–10 cells of *Aa* and *Pg,* and 10–100 cells of *Tf.* The carriage rates of *Aa* were relatively low (range 8–26%; mean 19%) and those of *Tf* relatively high (range 47–65%; mean 56%) regardless of the age cohort, whereas the rate of *Pg* increased from 12 to 57% due to ascending age cohorts up to 64 years, then decreased to 50%. None of the pathogens was found in 22%, a single pathogen in 50%, and multiple pathogens in 28% of the study population. Due to an increase in *Pg,* the carriage of multiple pathogens became more frequent by age, being highest (43%) in the cohort of 60–64 years. **Conclusion:** Multiplex PCR proved to be a sensitive and cost-effective method for simultaneous detection of *Aa, Pg,* and *Tf* directly from saliva. The examined periodontal pathogens had distinct carriage patterns: while *Aa* colonized the minority and *Tf* the majority of this adult population regardless of age, the salivary carriage rate of *Pg* steadily increased by age. Multiplex PCR is suitable for large-scale studies in order to evaluate the periodontal pathogen load in a population and the carriage as a risk factor for other diseases.

Supported by the Academy of Finland.

10

The Molecular Basis of Antibacterial Resistance in *Staphylococcus aureus*

E.E. Udo

Department of Microbiology, Faculty of Medicine, Kuwait University, Kuwait

Introduction: Antibacterial resistance is a major and increasing problem among *Staphylococcus aureus* isolates in hospitals and in the community because of their ability to acquire and rapidly transmit genetic elements. This study investigated the genetic location of antibacterial resistance and their transmission in methicillin-resistant *S. aureus* isolated in Kuwait hospitals. **Methods:** Antibacterial resistance was determined by disk diffusion or by measuring their minimum inhibitory concentrations. Plasmid isolation and genetic transfer experiments were performed according to standard protocols. The clonal relatedness of the isolates was determined by pulsed-field gel electrophoresis. **Results:** The isolates demonstrated resistance to multiple antibacterial agents, some of which varied over time but were all susceptible to vancomycin and teicoplanin. They contained plasmids of 1.9, 2.8, 3.5, 4.4, 26–28 and 38 kb in size. These plasmids could be transferred between *S. aureus* strains in conjugation or phage-mediated conjugation experiments. Results of the transfer experiments revealed that resistance to heavy metals, propamidine isethionate and ethidium bromide were linked and carried on the 26- to 28-kb plasmids. Mupirocin resistance was located on 32- to 40-kb plasmids, while chloramphenicol resistance was located on small 3- to 4.4-kb plasmids and no resistance phenotype was associated with the small 1.9- to 2.8-kb plasmids. Resistance to methicillin, gentamicin, kanamycin, tobramycin, streptomycin, tetracycline, trimethoprim, ciprofloxacin, fusidic acid and erythromycin were located on the bacterial chromosome. Pulsed-field gel electrophoresis demonstrated that few clones of methicillin-resistant *S. aureus* isolates circulated in Kuwait hospitals. **Conclusion:** This study has elucidated the genetic nature of antibacterial resistance and demonstrated that multiple antibacterial resistance is a major and increasing problem in methicillin-resistant *S. aureus* in Kuwait. It has also defined their clonal types.

Supported by Kuwait University Research Administration grants MI 098, MI 091.

11

Cytomegalovirus Genotypes and Drug-Resistant Mutants among Kidney Transplant Patients in Kuwait

N. Madi, W. Al-Nakib, A. Pacsa, A.S. Mustafa

Department of Microbiology, Faculty of Medicine, Kuwait University, Kuwait

Background: Cytomegalovirus (CMV) infection, in addition to being a major cause of morbidity and mortality in kidney transplant recipients, is associated with an immunosuppressive state and allograft rejection. This study focuses on the establishment and application of molecular methods to diagnose CMV infection in kidney transplant immunosuppressed patients in Kuwait, to measure viral load, to identify the different CMV genotypes prevalent in immunosuppressed patients and to investigate the phenomenon of drug resistance among these patients. **Methods:** PCR and real-time PCR were developed to detect and measure quantitatively CMV DNA in patients' blood, respectively. In addition, DNA sequencing of CMV UL97 and UL54 genes was performed to detect ganciclovir-resistant mutants and restriction fragment length polymorphism of gB and gH genes was used to genotype different CMV strains, as these have been associated with differences in CMV disease severity. **Results:** We were able to develop a PCR method with high specificity that can detect as little as 0.01 μg/ml (equivalent to 6 DNA copies/μl). Furthermore, we were able to show that we can quantify CMV DNA as few as 3 copies/μl using specific CMV probes by real-time PCR. Initial clinical testing using PCR showed that 58% of samples from immunosuppressed patients were positive for CMV in the blood. In addition, the data showed that 9 of 92 patients investigated, who had been given ganciclovir prophylactically, were still positive for CMV DNA by PCR, suggesting drug resistance. Data obtained from DNA sequencing will be presented to show the position of mutations in CMV drug-resistant mutants. Furthermore, our results showed that different gH and gB genotypes of CMV are circulating among the patients in Kuwait. **Conclusion:** Using the molecular techniques of PCR and real-time PCR, we were able to detect and quantify CMV DNA in the blood of a large number of kidney transplant patients. Early data suggest that CMV genotypes may influence the severity of the disease and drug resistance may be prevalent among these patients.

Supported by Kuwait University Administration Research grant YM 04/02 and College of Graduate Studies, Kuwait University.

12

Molecular Cloning, High-Level Expression, Purification and Immunological Characterization of Six Mammalian Cell Entry Proteins Encoded by *mce3* Operon of *Mycobacterium tuberculosis*

S. El-Shazly, Z. Khan, R. Al-Attiyah, A.S. Mustafa, S. Ahmad

Department of Microbiology, Faculty of Medicine, Kuwait University, Kuwait

Objective: To clone, express, and purify the six mammalian cell entry proteins encoded by *mce3* operon of *Mycobacterium tuberculosis* in *Escherichia coli* and to characterize the recombinantly expressed proteins for antibody reactivity using sera from tuberculosis (TB) patients. **Methods:** The pGES-TH-1 expression vector was used. The DNA segments corresponding to *mce3A, mce3B, mce3C, mce3D, mce3E* and *mce3F* genes were amplified by PCR using genomic DNA from *M. tuberculosis* as template, cloned and expressed in *E. coli* as fusion proteins with glutathione-*S*-transferase (GST) as the fusion partner at the N-terminal end and a 6 × His tag at the C-terminal end. The expressed proteins were detected by Coomassie blue staining of SDS-PAGE gels and characterized by Western immunoblotting. The recombinant fusion proteins were purified by affinity chromatography on Ni-NTA agarose, the fusion partner was cleaved by thrombin protease and the free Mce proteins were purified by preparative SDS-PAGE. Antibodies to all the six purified fusion proteins and the three pure Mce proteins (Mce3A, Mce3D and Mce3E) were raised in rabbits and characterized for B-cell epitopes using synthetic peptides. **Results:** With suitable manipulations, the DNA segments corresponding to all the six *mce3* genes *(mce3A–F)* of *M. tuberculosis* were successfully amplified by PCR, cloned and their identity was confirmed. The recombinant GST-Mce3 fusion proteins appeared as major cellular protein in Coomassie blue-stained SDS-PAGE gels, reacted with anti-GST antibodies, anti-penta-His antibodies and were purified by affinity chromatography on Ni-NTA-agarose. The Mce3A, Mce3D and Mce3E were also purified free of the fusion partner, the GST. All the six fusion proteins, particularly GST-Mce3A, GST-Mce3D and GST-Mce3E, reacted with human sera from TB patients. The antibodies raised against each fusion or pure Mce protein reacted with the corresponding Mce protein only, while anti-GST antibodies did not react with any Mce protein. Several B-cell epitopes were identified in Mce proteins by peptide mapping using synthetic peptides covering the entire lengths of Mce3A, Mce3D and Mce3E. **Conclusions:** The results of this study show that by suitably modifying the amplification protocols, the DNA segments, corresponding to families of important proteins encoded by *M. tuberculosis* genome, can be specifically amplified by PCR and expressed and purified. The purified proteins can also be used to raise antibodies in experimental animals. The recombinantly expressed proteins recognize antibodies in clinical samples from infected patients. The antibodies can also be used to detect in vivo expression of these proteins under various conditions.

Supported by Kuwait University Research Administration grant MI 05/00 and College of Graduate Studies, Kuwait University.

13

Synthetic Peptides Covering the Sequence of the Protein Encoded by the Mycobacterial MPB70 Gene Identify Naturally Derived Human T Cell Epitopes

R. Al-Attiyah[a], F. Shaban[a], H.G. Wiker[b], F. Oftung[c], A.S. Mustafa[a]

[a]Department of Microbiology, Faculty of Medicine, Kuwait University, Kuwait; [b]Division for Environmental Medicine and [c]Division for Infectious Diseases Control, Norwegian Institute of Public Health, Oslo, Norway

Introduction and Objectives: MPB70 is a major secreted protein antigen encoded by a gene common to *Mycobacterium bovis* and *Mycobacterium tuberculosis.* This study was conducted to identify the dominant natural Th1 cell epitopes of MPB70 in *M. bovis* BCG-vaccinated humans. **Materials and Methods:** Peripheral blood mononuclear cells (PBMC) from healthy subjects were tested with complex antigens, MPB70 and with a series of 13 overlapping synthetic peptides covering the entire sequence of MPB70 in proliferation and IFN-γ assays. Antigen-specific T cell lines were established from the healthy donors. The T cell lines were tested with MPB70 and its synthetic peptides in both the proliferation and IFN-γ assays. HLA restriction analysis was performed with anti-HLA class II-specific antibodies and HLA-typed antigen-presenting cells. Cytotoxicity assays were performed with antigen/peptide-pulsed monocytes/macrophages. **Results:** A total of 15 T cell lines specific for MPB70 were established from an equal number of donors. Testing of PBMC with the individual synthetic peptides revealed that the peptides p5, p6, p8, p12 and p13 were most frequently recognized in both assays. Moreover, the peptides from the carboxy terminus (p8, p12 and p13) were mostly recognized by the antigen-specific T cell lines in proliferation and/or IFN-γ assays. Other peptides stimulated the T cell lines from less than 5 donors. HLA restriction analysis showed that multiple HLA-DR molecules restricted the presentation of MPB70 as well as the peptides p12 and p13 to T cells. In addition, the responding T cell lines mediated antigen/peptide-specific cytotoxic activity against monocytes/macrophages pulsed with the whole protein and the peptides p8, p12 and p13. **Conclusion:** The results show that the peptides p12 and p13 constitute the dominant epitopes of MPB70, which are presented promiscuously to human Th1 cells of the protective phenotype. The promiscuous presentation of these peptides is further supported by the prediction that they can bind to a large number of HLA-DR molecules.

Supported by Kuwait Foundation for Advancement of Sciences (KFAS) grant No. 97-07-05.

14
Identification of *Mycobacterium tuberculosis*-Specific Proteins as Candidate Antigens for Diagnosis and Improved Vaccines against Tuberculosis

H.A. Amoudy[a], S. Ahmad[a], B. Robertson[b], J. Thole[b], D.B. Young[b], A.S. Mustafa[a]

[a]Department of Microbiology, Kuwait University; [b]Department of Infectious Diseases and Microbiology, Imperial College, University of London, UK

The worldwide increase in the incidence of tuberculosis (TB), the appearance of multidrug-resistant strains of *Mycobacterium tuberculosis* and the failure of Bacille Calmette-Guérin (BCG) vaccine to protect against TB have spurred the need for improved diagnostic tests and vaccines against a disease which is expected to account for more than 30 million deaths in the next decade. Recently, it has been discovered that a 9.5-kb genomic region designated RD1 in *M. tuberculosis* and *M. bovis* is deleted from BCG. The aim of this study was to investigate: (1) whether proteins encoded by these open reading frames (ORFs) represent dominant determinants in B and T cell responses to *M. tuberculosis* and (2) if these antigens could be used in diagnostic tests and improved vaccines against TB. Two ORFs (ORF11 and ORF14) were amplified from genomic DNA of *M. tuberculosis,* cloned in plasmid vectors and expressed in *Escherichia coli.* The expressed proteins were purified and analyzed for specificity and dominance in B and T cell recognition by comparing their reactivity to antibodies and T cells from TB patients and healthy BCG-vaccinated subjects. Purified ORF14 protein showed good antibody reactivity with sera from TB patients and it could discriminate between TB patients, their healthy contacts and healthy subjects in serologic tests. Both proteins, however, induced low reactivity and minimal production of INF-γ in in vitro T cell proliferation assays. In addition, synthetic peptides (25-mer overlapping with each other by 10 amino acids) covering entire sequences of RD1 ORFs, ORFs 2–11 and ORFs 14–15, were designed and used to study T cell responses of peripheral blood mononuclear cells from TB patients and healthy donors. The results showed that in TB patients, the strongest responses were obtained with the region covering the ORFs 5, 6, and 7. However, in healthy donors, ORFs 6 and 7 gave lower responses. In conclusion, ORF5 may be useful in developing new vaccine and ORFs 6 and 7 in the diagnosis of TB.

Supported by a grant from the College of Graduate Studies, Kuwait University, Kuwait.

15
Cloning and Expression of a Cytotoxin Gene and Characterization of the Cytotoxin from *Campylobacter jejuni*

M.J. Albert[a], I. Khan[a], B. Adler[b]

[a]Faculty of Medicine, Kuwait University, Kuwait; [b]Department of Microbiology, Monash University, Melbourne, Australia

Campylobacter jejuni is a major foodborne pathogen and a leading cause of diarrhoea worldwide. A cytotoxin is most likely involved in the pathogenesis of inflammatory diarrhoea due to *C. jejuni.* A 45-kDa outer membrane protein encoded by the *porA* gene was reported to exhibit cytotoxic property for cultured mammalian cells in vitro. The *porA* gene is an open reading frame of 1,275 bases. Our objective was to clone the *porA* gene, express it in a suitable expression vector and characterize the expressed protein for its functional properties. We amplified the *porA* gene (also referred to as *ctt,* for cytotoxin gene) from a cytotoxic strain of *C. jejuni* by PCR. It was cloned and expressed in the *Escherichia coli* codon plus RIL strain using the fusion vector, pGEX-4T-1. The fusion protein was present as inclusion bodies. In preliminary studies, the fusion protein did not exhibit cytotoxic property for Chinese hamster ovary cells or induce diarrhoea in a rat ileal loop assay. Attempts are being made to solubilize the protein and cleave it with thrombin and study the functional properties of the cleaved protein. This will reveal whether or not the *porA* gene product is indeed the cytotoxin. This will help us to prepare knockout mutants defective for the *porA* gene and study their pathogenic potential. Sixty clinical isolates of *Campylobacter* spp. from Kuwait were screened for the *ctt* gene by PCR and 47 were found to be positive. This suggested a high prevalence of the cytotoxin gene among the Kuwaiti isolates.

Supported by the Kuwait University Research Administration grant No. MI 04/00.

16
Rapid Diagnosis of Enteroviral Meningitis and Genotyping of Clinical Isolates of Enteroviruses in Kuwait, Using RT-PCR Followed by DNA Sequencing

A. Dalwai, S. Ahmad, S. George, A. Pacsa, E. Sadek, W. Al-Nakib

Department of Microbiology, Faculty of Medicine, Kuwait University, Kuwait

Objectives: Enteroviruses are important causative agents of aseptic meningitis, myocarditis and type I diabetes. The aim of this study was to establish a sensitive method for the rapid diagnosis of aseptic meningitis cases by detection of enteroviral RNA in cerebrospinal fluid (CSF) using RT-PCR and to genotype the enterovirus strains prevalent in Kuwait. **Methods:** Reference enterovirus strains were obtained from American Type Culture Collection (ATCC), Rockeville, Md., USA, and grown on Vero cell lines. Consensus and enterovirus-specific primers based on CLUSTAL W alignment of more than 30 different enteroviral sequences were used for the amplification of enteroviral RNA in RT-PCR. CSF had been obtained from patients suspected of aseptic meningitis since the year 2002. The enteroviral RNA was extracted from reference enteroviruses and CSF samples by using a Qiagen RNA isolation kit, and used as template for the amplification reaction. The RT-PCR amplification was performed in a single tube as a two-step reaction, followed by semi-nested PCR and the products were detected by agarose gel electrophoresis. The amplified product was further confirmed by Southern hybridization with a enterovirus-specific biotinylated probe. The genotyping was performed by direct cDNA sequencing of the 5′ untranslated region (UTR) of the enteroviral genome. **Results:** The RT-PCR assay successfully amplified a fragment of expected size from all the ATCC enteroviruses that were tested. RT-PCR followed by semi-nested PCR was able to detect at least 10 copies of enteroviral RNA. A similar sensitivity was obtained for RNA extracted from spiked CSF specimens. No amplification was obtained when RNA or DNA prepared from other viruses, including rhinoviruses, was used as template. The specificity of detecting enteroviral RNA was further demonstrated by Southern hybridization using an enterovirus-specific probe. The established methodology, when applied to CSF samples obtained from suspected cases of aseptic meningitis, detected enteroviral RNA in 48 (26%) of 191 samples. Coxsackie viruses were the predominant enterovirus type detected in these cases as revealed by direct DNA sequencing of the 5′ UTR of the enteroviral genome. **Conclusions:** The established RT-PCR assay demonstrated its diagnostic utility when the method detected enteroviral RNA in 26% of CSF samples obtained from patients suspected of aseptic meningitis. Among enteroviruses, Coxsackie viruses were more frequently found as identified by direct DNA sequencing of the 5′ UTR of the enteroviral genome. This demonstrates the importance of enteroviruses as etiologic agents in causing aseptic meningitis.

Supported by Kuwait University Research Administration grants MI 04/01, YM 03/02 and College of Graduate Studies, Kuwait University, Kuwait.

Cancer

17
Molecular Basis of Cancer and Its Clinical Applications

C.H.J. Ford

Department of Surgery, Faculty of Medicine,
Kuwait University, Kuwait

There is a genetic component to cancer and a multistage progression to malignancy, although each of the stages cannot usually be clearly defined in a particular tumour. Current evidence suggests that most cancers arise from a single cell which has acquired some heritable form of growth advantage and further mutational events, involving genes responsible for control of cell growth, lead to tumour cell progression. Additional changes allow the outgrowth of clones with metastatic potential. The use of modern molecular techniques to study cancer over the last 2 decades has led to the identification of 4 major groups of genes which are involved in tumourigenesis – oncogenes, tumour suppressor (TS), cell cycle control (CCC) and mismatch repair (MMR). Cellular oncogenes (protooncogenes) encode proteins which are important in the control of cell proliferation, differentiation, CCC and apoptosis. Mutations in these genes act dominantly and lead to a gain of function. In contrast, TS genes inhibit cell proliferation by arresting progression through the cell cycle and block differentiation. CCC genes are involved in the positive and negative regulation of the cell cycle and they interact with oncogenes and TS genes, and in some cases may be considered to be such in their own right. To ensure that DNA replication is complete and that any damaged DNA is repaired, cells must pass through specific checkpoints and MMR genes ensure that damaged DNA is repaired. Oncogenes, TS, CCC and MMR genes are being used in classifying, determining diagnosis and prognosis, and as therapeutic targets. Identification of some cancer susceptibility genes has helped to improve the management of patients in families with a genetic predisposition to cancer by offering presymptomatic testing and hence earlier intervention and treatment of the disease in, for example, familial adenomatous polyposis and hereditary non-polyposis colon cancer, and familial breast and breast/ovarian cancer. The ability to detect the product of a *BCR/ABL* fusion gene in chronic myeloid leukaemia by PCR has resulted in a very sensitive diagnostic test for chronic myeloid leukaemia. Also, the use of real-time PCR to monitor the kinetics of residual *BCR-ABL* transcripts over time has shown that those patients who relapse have reappearance and/or rising levels of transcripts. Expression profiling using micro-array technology is being used to refine the classification of cancer with the greatest advances being seen in leukaemia, lymphoma and breast cancer. Perhaps the greatest impact that a better knowledge of cancer at the molecular level has had is the new avenues that have opened up for therapy. These include targeting receptors, TS and CCC gene products and angiogenesis molecules as well as immunologic intervention. Our understanding of the mechanisms involved in the development of cancer and in its treatment are being completely revolutionised and this should bring prevention of cancer and the ultimate goal of cancer therapy – the elimination or inactivation of each and every remaining malignant cell from the patient – considerably closer.

18
Hereditary Non-Polyposis Colorectal Cancer: Identification of a Novel *hMSH2* Gene Mutation in an Arab Family

M. Marafie[a], M. Motawy[b], F. Macdonald[c], J. Bell[c], S. Al-Awadi[a]

[a]Centre for Medical Genetics, and [b]Kuwait Cancer Control Centre, Kuwait; [c]West Midlands Regional Genetics Laboratory, Birmingham, UK

Hereditary non-polyposis colorectal cancer (HNPCC) is a highly penetrant autosomal dominant syndrome characterized by early-onset colon cancer and a variety of extracolonic tumours involving the endometrium, ovary, stomach, pancreas, small intestine, liver and urinary tract. HNPCC accounts for 5% of all colorectal cancers and germ line mutations in the DNA mismatch repair genes (MMR) are responsible for this disorder. Several MMR genes *(hMLH1, hMSH2, hMSH6, hPMS1, hPMS2)* have been identified. The majority (90%) of mutations have been reported in the *hMLH1* and *hMSH2* genes and only 10% in *MSH6* gene. We offered genetic counseling to a cohort of patients attending the familial cancer clinics at both the Kuwait Medical Genetic Centre and Cancer Control Centre, aiming to identify families with hereditary cancers. We report a highly consanguineous Arab family from Kuwait that fits the HNPCC syndrome. Pedigree analysis revealed the presence of 23 members of consecutive generations expressing colonic and extracolonic tumours, affecting both sexes and at different young ages. DNA analysis was offered to 3 members, the proband who developed colon cancer at age 37 years, her father who had colon cancer and her maternal aunt who developed endometrial cancer. Mutational analysis of all exons of the *hMSH2* gene was carried out by the PCR/SSCP test followed by direct sequencing of the fragment with band shift. Sequencing showed a germ line nonsense mutation Q76X in exon 2 of the *hMSH2* gene. Other members of the family are currently being investigated. Genetic counseling and screening for germ line mutations in individuals of such genetically predisposed families will help in determining those patients and those high-risk non-symptomatic mutation carriers who are likely to develop tumours.

19

Investigation of Epstein Barr Virus in Gastrectomy Specimens

Y.A. Luqmani[a], *S.O. Linjawi*[b], *S. Shousha*[b]

[a]Department of Pharmaceutical Chemistry, Faculty of Pharmacy, Kuwait University, Kuwait;
[b]Charing Cross Hospital, Imperial College of Medicine, University of London, London, UK

The presence of Epstein Barr virus has been reported in a small proportion of gastric carcinomas in several populations, particularly in East Asia, where this cancer is prevalent. It has been suggested that it may play a role in the pathogenesis of cancer in these patients, but the origins of the infection have not been identified. In this study, we examined 43 consecutive gastrectomy specimens containing a variety of benign and malignant lesions. In situ hybridisation was used to detect EBER RNA, a marker of latent viral infection. Only 1/20 (5%) gastric cancers was found to be EBER+; a moderately differentiated adenocarcinoma with heavy lymphocytic infiltration. Double immunostaining with antisera against B and T cell markers indicated that the interstitial lymphocytic infiltrate was predominantly of the B cell type, but the majority of lymphocytes overlying the tumour cells were CD8+ T cells. The other gastric lesions examined, which included 15 peptic ulcers, 6 stromal tumours and 2 lymphomas, were all negative for EBER. Using a biotin detection system, scattered EBER+ cells were observed in adjacent normal gastric and/or duodenal mucosa in 9 sections from 8 cases (19% of the 43 patients examined). However, this reactivity, in normal cells, was not seen when digoxigenin was used for detection. The apparently EBER+ cells also showed immunoreactivity with an antiserum against chromogranin A, suggesting that these were neuro-endocrine cells within the gastric and duodenal mucosa. It was concluded that Epstein Barr virus infection was restricted to the lympho-epithelioma type of gastric tumour and was absent from the other adenocarcinomas. The occasional EBER+ signal observed over non-cancerous cells was likely due to an anomalous reaction from neuro-endocrine cells and was indeed an artifact of the biotin detection system.

Supported by grants from the Gunnar Nilsson Research Trust UK and Dako Ltd. to S.S.

20

Glycolaldehyde-Induced DNA Damage Associated with Apoptosis in Human Breast Carcinoma

M. Al-Maghrebi[a], *F. Al-Mulla*[b], *L.T. Benov*[a]

Departments of [a]Biochemistry and [b]Pathology, Faculty of Medicine, Kuwait University, Kuwait

Introduction: The accumulation of Maillard reaction products in tissue proteins is implicated in the development of pathology in aging, diabetes and atherosclerosis. Intermediates of this reaction include highly reactive compounds such as methylglyoxal, glyoxal, and glycolaldehyde. Accumulation of these metabolites to toxic levels could affect various cellular functions. Despite the growing interest in these short chain fragments, their role in cell damage is still not fully recognized. **Objectives:** This study seeks to determine the effect of glycolaldehyde on cell metabolism and proliferation in breast carcinoma cell line T47D. **Methodologies:** The effect of glycolaldehyde on cell growth and viability was evaluated by trypan blue exclusion. At definite time intervals, membrane-free lysates were used for assaying enzyme activities. DNA damage was detected by labeling the 3′-OH termini of DNA molecules using terminal deoxynucleotidyl transferase-mediated dUTP nick end labeling. **Results:** We found that glycolaldehyde exerts a cytostatic and cytotoxic effect on breast carcinoma cell line T47D. The cytostatic effect appears to be due to inhibition of key metabolic enzymes. The cytotoxic effect was shown by DNA fragmentation caused by the production of superoxide radical from glycolaldehyde autoxidation. DNA fragmentation was assessed by the increased number of labeled DNA ends after treatment with glycoladehyde. A cell-permeable superoxide dismutase mimetic prevented the cytotoxic effect of glycolaldehyde. **Conclusion:** Glycolaldehyde induces DNA strand breaks and inhibition of key metabolic enzymes leading to apoptosis in breast carcinoma cell line T47D.

Supported by Kuwait University Research Administration grants MB031 and MB01/01 and the Kuwait Foundation for the Advancement of Science (KFAS) grant 990707.

21

Genome Instability and Colorectal Cancer: Implications for Life and Death of the Cancer Cell

A.H. Wyllie

Department of Pathology, University of Cambridge, Cambridge, UK

About 60% of sporadic colorectal cancers show instability in the number and structure of chromosomes (CIN). There are many trisomies and invariably unbalanced translocations, although these do not appear to cluster in any particular chromosome locations. The changes represent a true instability, and not the establishment of a new stable state, following a period of disruption or damage. A further 10–15% of tumours show unstable microsatellite sequences (MIN). In the overwhelming majority of these, this is due to epigenetic silencing of the mismatch repair genes $hMLH_1$ and (less commonly) $hMSH_2$ by methylation. These tumours usually show stable chromosomes. A third set, accounting for the remaining 25–30% of sporadic tumours, has stable microsatellites and minimal chromosome changes only (MACS). This crude classification into CIN, MIN and MACS cancers may be subject to further subdivision on the basis of more detailed molecular analysis, but even at present has implications for prognosis and – perhaps – treatment resistance and sensitivity. There are theoretical reasons why this should be so: experiments with mice bearing germ line defects in capacity for mismatch repair or the detection of double-strand DNA breaks in chromosomes show differential losses of capacity to activate apoptosis after various treatments. Little is known of the biology or therapeutic responses of the MACS subgroup.

22

Carbonyl Reductase: Tales of Mice and Women

F. Al-Mulla [a], *M. Umemoto* [b], *Y. Yokoyama* [b], *S. Sato* [b], *S. Tsuchida* [c], *Y. Saito* [b]

[a]Department of Pathology, University of Kuwait, Kuwait;
[b]Department of Obstetrics and Gynecology, and
[c]Second Department of Biochemistry, Hirosaki University School of Medicine, Hirosaki, Japan

To explore reasons for differences in the malignancy of tumors, we have compared two cell lines derived from a mouse lung adenocarcinoma cell line that differ 10-fold in their capacity to form lung metastases from subcutaneous primary tumors or after intravenous injection. One mRNA encoding carbonyl reductase (CR) was identified at a relatively high abundance in the subline with low metastatic capacity but was not detectable in the highly metastatic subline. Transfection of the former subline with a plasmid construct expressing antisense CR rendered the cells highly metastatic. Conversely, the capacity of the highly metastatic cells to metastasize was markedly reduced after transfection with a construct expressing CR. As there are currently no data addressing a similar function for CR in human cancers, the aim of this study was to assess the correlation between survival and metastasis, and the CR level in epithelial ovarian cancer. Using anti-CR antibody, immunohistochemical staining was performed on 73 epithelial ovarian cancers, 13 borderline malignant tumors, and 25 benign ovarian tumors for a total of 111 specimens. The combined rate for strongly and weakly positive reactions for CR was 32.0% for benign tumors, 38.5% for borderline malignant tumors, and 61.6% for ovarian cancers. The CR-positive rate was 35.7% (weakly positive alone) for ovarian cancers with retroperitoneal lymph node (RLN) metastasis and 67.8% for those without RLN metastasis ($p < 0.05$). The 5-year survival rate was 62.7% for the patients with CR-negative cancer and 86.1% for those with CR-positive cancer ($p < 0.05$). The present results indicate that decreased CR expression in epithelial ovarian cancer is associated with RLN metastasis and poor survival.

Fahd Al-Mulla was supported by the Kuwait Foundation for the Advancement of Sciences grant No. 990707.

23

Expression of G1-S Cyclins in Multistage Human Pancreatic Duct Cell Carcinogenesis

M.A. Al-Aynati, J. Ho, M.-S. Tsao

Department of Pathology, Princess Margaret Hospital, University Health Network, and Department of Laboratory Medicine and Pathobiology, Faculty of Medicine, University of Toronto, Toronto, Canada

Cell cycle deregulation is an important feature of cancer cell biology, but the expression of different cyclins that regulate the G1-S cell cycle checkpoints during early stages of multistep pancreatic carcinogenesis has not been comprehensively investigated. We constructed tissue microarray (TMA) blocks from 94 Whipple resection specimens. We used immunohistochemistry to analyze the expression of cyclins D1, D3, E and A proteins in pancreatic intraepithelial neoplastic (PanIN) duct lesions and contrasted their expression patterns with those in normal ducts and ductal cancers. Cyclin D3 protein overexpression was noted early in PanIN-1 A lesions, as well as in practically all the PanIN-2 lesions. Cyclins E and A overexpression occurred initially in PanIN-2 lesions and the prevalence peaked in cancer. Cyclin D1 did not show a significantly differential expression in normal ducts, PanIN lesions and adenocarcinoma. This study demonstrates the utility of TMA for comprehensive studies of gene expression in PanIN lesions. The results have also revealed the potentially significant role that cyclin D3 plays in deregulating proliferation during the early stages of pancreatic duct cell carcinogenesis.

Supported by the Canadian Institutes of Health Research grant MOP-49585.

24

A Novel Germ Line Mutation in the von Hippel-Lindau Gene in Kuwaiti von Hippel-Lindau Patients

S. Al Fadhli

Department of Medical Laboratory Sciences, Faculty of Allied Health Sciences, Kuwait University, Kuwait

Objective: We studied the germ line mutation in an extended family in which one member was diagnosed clinically with von Hippel-Lindau (VHL) disease. **Materials and Method:** In this study, we used PCR/SSCP sequencing to identify the germ line mutation of the *VHL* gene in the patient as well as to screen 9 other members of the family. **Results:** The patient and 3 other members of the family were shown to have the same mutation in the splice donor site of the first intron. The mutation was identified as IVS1+1G→T. **Conclusion:** This early detection of *VHL* gene mutation will definitely help in the early diagnosis and proper genetic counseling of these mutation carrier patients in this family. In addition, screening more family members is planed.

Supported by the Kuwait University Research Administration grant No. NM00/01.

25
Molecular Cytogenetic Analysis of Breast Tumour Samples

M.A. Alwohhaib, J. Parry, L. Tammim

Department of Medical Laboratory Science, Faculty of Allied Health Sciences and Nursing, Kuwait University, Kuwait

Comparative genomic hybridization (CGH) provides a genome-wide search for the chromosomal regions that are either amplified or deleted during the somatic cancer development process. A large number of tumour analysis studies using CGH have been published in recent years, indicating the efficiency of the technique. In addition, there are a number of regions of abnormality that have been discovered by CGH, such as the androgen receptor gene, located in a region of recurrent abnormality at Xq12 in hormone-refractory prostate cancers, and PIK3CA, located in a region of recurrent abnormality at 3q26 in ovarian cancer. A large number of women between 31 and 40 years of age have been diagnosed with breast cancer at the Kuwait Cancer Control Centre (KCCC). CGH was used to look for the copy number changes in the genome of 21 Kuwaiti women who are under 40 years of age and who had been diagnosed with primary cancer at the KCCC. Gain in chromosomal regions 4p (48%), 8p (52%), 10p (48%), 12p (62%), 13q (48%) and 21q (48%) was detected. In summary, because of the advantages of the CGH technique, it was used to identify and map increases as well as decreases in DNA sequence copy number in 21 frozen breast tumours, which were taken from Kuwaiti women who were under 40 years of age. It is hoped that the results may lead to an improved understanding of carcinogenesis and its progression.

26
Characteristics of Male Breast Cancer in Kuwait

Y.A. Luqmani

Department of Pharmaceutical Chemistry,
Faculty of Pharmacy, Kuwait University, Kuwait

Male breast cancer is a relatively rare occurrence, representing 1% of all male malignancies, as a consequence of which its biology and aetiology remain poorly understood. Treatment is somewhat empirically applied using experience from treatment of the cancer in females. In this study, case notes were examined of 41 patients (aged 25–82 years) registered in Kuwait hospitals between 1979 and 1998. Five-year actuarial rates were 58 and 67%, respectively, for disease-free and overall survival. Only grade and node status achieved prognostic value in multivariate analysis. Available archival pathology material from 18 of these patients was analysed by immunohistochemistry for altered expression of several proteins associated with the disease affecting females. Tumour cells exhibited increased reaction to antisera against oestrogen receptor (ER; in 89% of cases), progesterone receptor (PR; 61%), pS2 (44%), cathepsin D (72%), c-erbB2 (50%), p53 (56%), bcl-2 (78%) and Ki67/PCNA (100%). Bivariate analyses indicated associations between some of these factors, though none influenced survival. Genetic analysis of 25 microsatellite repeat loci showed repeat number instability in 5 patients; at a single site in 1 and at 3, 7 and 9 sites in another 3 cases (most frequently at D2S441 and D13S325). Loss of specific heterozygosity was apparent in 8 patients; at multiple sites in 4 of these. Thus, replication error and allelic loss characterise male breast cancer in much the same way as in women. These observations indicate, in general, clinicopathological similarities with the more prevalent female cancer, but with some differences in expression of endocrine and other factors which are relevant for more rational approaches to treatment. Relatively frequent occurrence and high level of ER has justified the use of anti-oestrogen therapy but without prima evidence of involvement in proliferation, as reflected by a lack of correlation with inducible factors such as PR and cathepsin D. The greater frequency of c-erbB2 positivity than is usual in women also suggests that antagonism of proliferative pathways involving the epidermal growth factor receptor family may prove a more effective therapeutic strategy than attempting to block ER.

Supported by Kuwait University Research Administration Grants NPM 032 and NPM 042.

27
Novel Approaches to Genetic Diseases

P.T. Ozand

Department of Genetics, King Faisal Specialist Hospital and Research Centre, Riyadh, Saudi Arabia

The objective of this presentation is to present the novel strategies adopted to prevent genetic diseases in Saudi Arabia. The genetic diseases either due to chromosome defects or to alterations in single gene structure produce infants with usually severe diseases. The management of such disorders is: (a) usually costly due to the fact that the morbidity lasts for many years, (b) not always successful and (c) disrupts the normal family function. These diseases, particularly those inherited single-gene disorders, plague the communities with consanguineous marriages. Based on our experience in the Kingdom, we have adopted several strategies to combat this public health problem. The methods to combat genetic diseases include: (1) if nothing else is possible to perform a neonatal screening program for treatable diseases. This must be done within 2–3 days after birth. (2) Preimplantation diagnosis. Initially, we have focused on six major single-gene diseases of the country: maple syrup urine disease, biopterin-dependent phenylketonuria, homocystinuria, propionic acidemia, Niemann-Pick disease type B and Gaucher disease type A. All of these diseases are either difficult to manage, requiring the devotion of many clinical hours by the physician or their procedures of management, or are extremely costly. (3) Premarital screening. Applying the mutations on a DNA chip and screening the extended family as well as population for carriers of these diseases. (4) Chromosome abnormalities to be studied by complete human genome screening that can be applied to a newborn as well as in preimplantation efforts. This presentation will detail these approaches and our preliminary results. The result with preimplantation intervention in an ataxia-telangiectasia family will be presented.

28

Mutations Underlying Metabolic Diseases in Patients from Kuwait: Phenylketonuria, Carbonic Anhydrase II Deficiency and Tay-Sachs Disease

E. Samilchuk

Kuwait Medical Genetics Center, Kuwait

A number of metabolic diseases have been reported from Kuwait. Although being relatively common as a group, each particular metabolic disease is rare and often confined just to one or two families. Knowledge of mutations running in families is essential for proper genetic counseling because in most cases biochemical and enzymatic tests are not informative for carrier detection. Preimplantation and early prenatal diagnosis is also based on the direct testing for the mutation. Each disease requires mutation analysis of a specific gene, which is further complicated by the ethnic variability of mutations. DNA samples collected from 13 unrelated phenylketonuria (PKU) families, 1 family with carbonic anhydrase II deficiency (CAII-D), and 3 unrelated Tay-Sachs disease (TSD) families were subjected to mutation analysis of PAH, CAII and HEXA genes, respectively. The major techniques used were polymerase chain reaction, restriction fragment length polymorphism and heteroduplex analysis, and DNA sequencing. The analysis revealed PAH mutations in all PKU families (26 alleles in total): 1066nt11g→a in 34.6%, K363fsdelG in 15.4%, IVS4nt5g→t in 15.4%, IVS2nt5g→t in 11.5%, G352fsdelG in 7.6%, and P281L, E280K, I224T each in 3.8%. The 1066nt11g→a, the most frequent variant in Kuwait, is also the most common PKU mutation in many Mediterranean populations. In contrast, the IVS4nt5g→t and I224T seem to be rare mutations. All affected members from the CAII-D family were homozygous for the IVS2nt1a→g mutation of the CAII gene. This mutation has previously been reported in patients of Arabic origin from North Africa and Saudi Arabia. The R504H was found in 1 of the 3 TSD families. All affected members were homozygous for this mutation. The R504H has previously been reported in 3 patients (Lebanese, Assyrian and American). However, in 2 TSD families from Kuwait, the mutations remain unknown, despite the fact that all exons and splicing sites of the HEXA gene were sequenced. In conclusion, as a result of this study, carrier detection, prenatal, and preimplantation diagnosis is now possible in all families where the mutations have been identified.

29

Diabetic Neuropathy: Past, Present and Future

M.S. Bitar

Department of Pharmacology and Toxicology,
Faculty of Medicine, Kuwait University, Kuwait

Diabetes mellitus is a leading cause of death and it is associated with a number of complications including retinopathy, nephropathy, vasculopathy and neuropathy. Each has been linked to the severity of hyperglycaemia; thus, the mainstay of treatment has been to aggressively control serum glucose levels. This practice is supported by the large NIH-sponsored diabetes complication and control trials that linked control of glucose to delayed onset of complications. Other mechanisms may also contribute because restoration of glycaemia does not always abrogate progression of established disease. Thus, alternative therapeutic approaches based on an understanding of the mechanisms of diabetes-induced end-organ damage are needed. Peripheral neuropathy is one of the most common and disabling long-term consequences of diabetes mellitus. A failure of neurotrophic factor to regulate neuronal phenotype might be expected to result in such a clinical picture. Experimentally, diabetic subjects show reduced expression of target-derived insulin-like growth factor (IGF-1) as well as reduced expression of neuronal genes that are responsive to IGF-1. The latter is corrected by administration of exogenous IGF-1. Accordingly, deficient neurotrophic support might contribute to the pathogenesis of diabetic neuropathy. Moreover, future treatment of this pathology may include exogenous neurotrophics or other strategies to correct their deficiency of action.

Author Index

(A) = Abstract

Subject Index

KARGER
Fax +41 61 306 12 34
E-Mail karger@karger.ch
www.karger.com

Accessible online at:
www.karger.com/mmp

A novel focus on molecular approaches to neurological diseases

Principles of Molecular Neurosurgery

Editors
Andrew Freese
Frederick A. Simeone
Paola Leone
Christopher Janson

Neurosurgery; Neurology; Metabolic Diseases, Molecular Biology, Neurobiology, Oncology, Orthopedics

Table of Contents

www.karger.com/prnes

KI05432

Progress in Neurological Surgery, Vol. 18
Series Editor: L. Dade Lunsford (Pittsburgh, Pa.)
ISSN 0079–6492

Principles of Molecular Neurosurgery
Editors: Andrew Freese (Minneapolis, Minn.); Frederick A. Simeone (Philadelphia, Pa.); Paola Leone; Christopher Janson (Camden, N.J.)
XIV + 660 p., 96 fig., 12 in color, 25 tab., hard cover, 2005
CHF 368.– / EUR 263.– / USD 334.75
Prices subject to change
EUR price for Germany, USD price for USA only
ISBN 3–8055–7784–2

KARGER

Order Form

Please send: ______ copy/ies

Postage and handling free with prepayment

Payment:
☐ Check enclosed ☐ Please bill me

Please charge to my credit card
☐ American Express ☐ Diners ☐ Eurocard
☐ MasterCard ☐ Visa

Card No.: ____________________

Exp. date: ____________________

Orders may be placed with any bookshop, subscription agency, directly with the publisher or through a Karger distributor.

Fax: +41 61 306 12 34

S. Karger AG, P.O. Box, CH–4009 Basel (Switzerland)
E-Mail orders@karger.ch, **www.karger.com**

Name/Address: ____________________

____

____

____

____

Date: ____________________

Signature: ____________________

From the patient to the lab and back again

New Journal

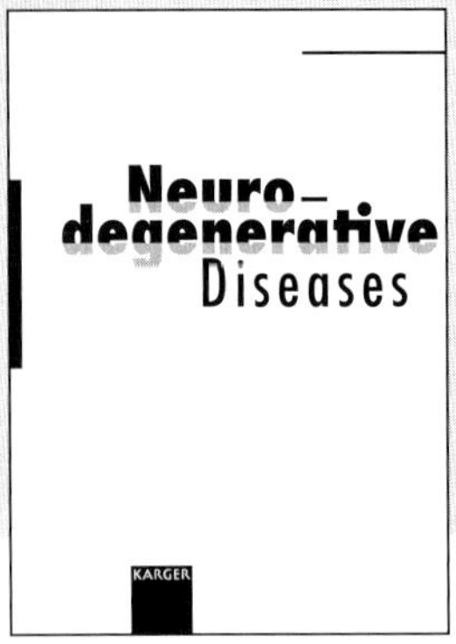

Neuro–degenerative Diseases

Editors-in-Chief
R.M. Nitsch, *Zürich*
C. Hock, *Zürich*

Neurodegenerative Diseases publishes results from basic and clinical scientific research programs designed to better understand the normal functions of genes and proteins involved in neurodegenerative diseases, to characterize their role in pathogenic disease mechanisms, to model their functions in animals and to explore their roles in the diagnosis, treatment and prevention of disease. This high-quality, multidisciplinary journal for the publication of advances in the understanding of neurodegenerative diseases, including Alzheimer's disease, Parkinson's disease, amyotrophic lateral sclerosis, Huntington's disease and related neurological and psychiatric disorders is coming soon.

Invitation for papers

Only original papers written in English are considered and should be sent to:

S. Karger AG
Editorial Office
Neurodegenerative Diseases
P.O. Box
CH–4009 Basel (Switzerland)
E-Mail neurodis@karger.ch

Mailing address for courier deliveries only:
Allschwilerstr. 10
CH–4055 Basel (Switzerland)

Sections:

Neurobiology
Molecular Biology
Cell Biology
Animal Models
Structural Biology

Clinical Research
Genetics
Neuropathology
Imaging
Biological Markers
Therapy

Drug Discovery
Biotechnology
Pharmaceuticals

Read it online:
www.karger.com/neurodis

Subscription rates per year

Vol. 1 free

Personal rates per year	Online	Print		Print/Online combined
	Subscription	Subscriptio…	…and print/online	Subscription
	CHF 3…		CHF 31.80 (Europe) CHF 47.10 (Overseas)	CHF 341.60
	EUR 2…	EUR 222.60	EUR 22.80	EUR 252.60
	USD 240.60	USD 240.60	USD 31.80	USD 273.60

Institutional rates per year	Online	Print		Print/Online combined
	Subscription	Subscription	Postage and handling added to print and print/online	Subscription
	CHF 1002.–	CHF 1002.–	CHF 39.60 (Europe) CHF 58.80 (Overseas)	CHF 1043.–
	EUR 742.–	EUR 742.–	EUR 28.50	EUR 772.–
	USD 802.00	USD 802.00	USD 39.60	USD 835.00

Airmail surcharge per year: CHF 39.60 / USD 33.00
Prices subject to change
EUR price for Germany, USD price for USA only

Neurodegenerative Diseases
2004: Volume 1
with 6 issues
Language: English
ISSN 1660–2854
ISSN-Online 1660–2862

KF 04050

KARGER